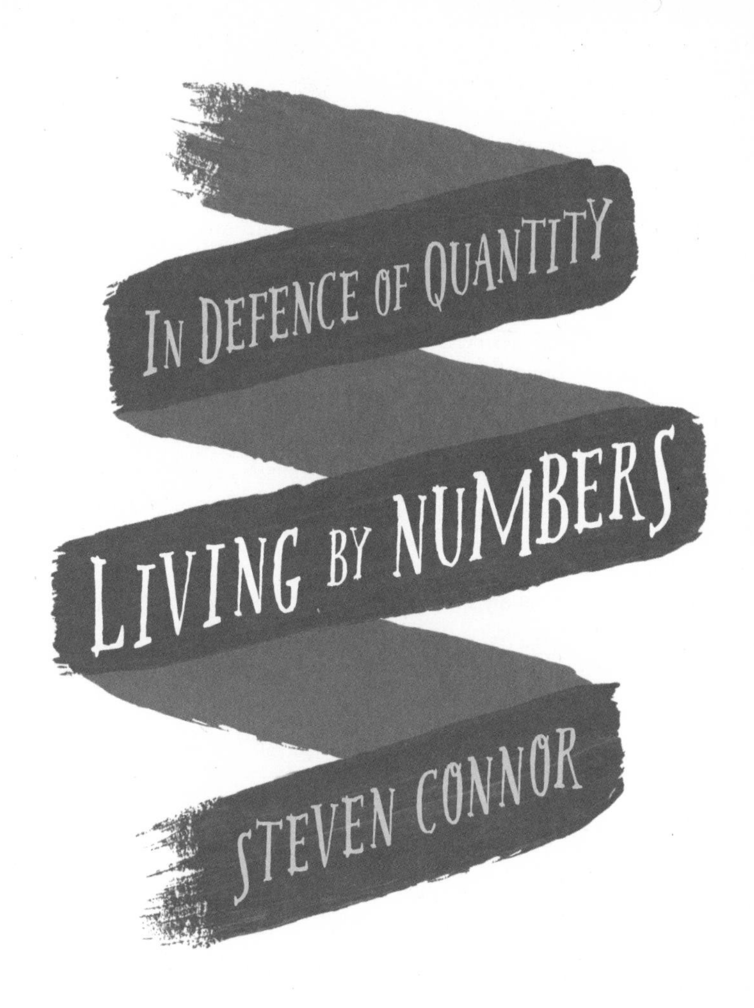

REAKTION BOOKS

Published by Reaktion Books Ltd
Unit 32, Waterside
44–48 Wharf Road
London N1 7UX, UK

www.reaktionbooks.co.uk

First published 2016
First published in paperback 2017

Printed and bound in Great Britain
by Bell & Bain, Glasgow

A catalogue record for this book is available from the British Library

ISBN 978 1 78023 825 8

CONTENTS

1

VERNACULAR MATHEMATICS

'There's nothing you can't turn into a sum, for there's nothing but what's got number in it.'[1]
Bartle Massey in George Eliot, *Adam Bede* (1859)

Sooner or later, in writing anything, be it novel, memoir, essay, annual report or sonnet, one will have to write the words that will stand, or be allowed to stand, as the first, whether or not these happen to be words like 'Chapter One' which explicitly start the count. There are all kinds of ways in which writers can lament, defy, denounce, ignore or attempt to evade this ordinality, but all of them testify to its inescapable force, which can be transacted with, but never transcended. All acts of writing will have, not only to have beginnings, but to have been begun. In any ordinal proceeding, any opening sentence, or opening word of such a sentence, begins a count, an order of things.

The fact that words have to be in some order or another is just one of the simplest ways in which verbal and numerical signs are commingled. Along the way, this book aims to consider some of the many other forms in which words, signs and numbers work on and with each other. It is prompted by what seems to be a striking contemporary contradiction. On the one hand, our lives are more than ever penetrated by and even perhaps governed by number, in all its aspects. On the other, it seems to many, and perhaps especially those who think of themselves as participating in the world of arts, culture or the humanities, that number has come to be a kind of tyrant virus in human affairs, such that the preservation of our humanity, not to mention its earthly vicar, 'the humanities', depends upon our recession from number, and the rescue of what are thought to be qualities from their reduction to mere quantities. The strength and prevalence of this anti-numerical animus is remarkable, not least because there has never been a time in human history in which it has been less possible to distinguish quantitative thinking and living from qualitative. Just at the moment, in other words, at which the commingling of quality and quantity has become both unignorable and inextricable, an anti-numerical ideology founded on the principle that it should be possible to separate quality from quantity maintains itself with unexampled vigour and self-reproducing virulence.

Later chapters in this book will consider the motives, forms and effects of this number-phobia, while also trying to show that, far from representing the inhuman, numbers are intrinsic to almost all conceivable kinds of human being. We ought perhaps to think of our relation to numbers in the way suggested by the old man Chremes in Terence's play *Heauton timorumenos (The Self-tormentor)* who declares: 'Homo sum: humani nihil a me alienum puto.' 'I am a man: I think nothing human alien to me.' Numbers may seem inhuman and yet the making out of numbers and numerical relations seems one of the most identifiably and irrepressibly human of accomplishments. The question is perhaps whether numbers are alien or not to human beings – including perhaps the question of whether that 'nihil' in Terence's phrase is to be considered a number, or itself a kind of alien principle within number. Chremes' words, admired, among many others, by Augustine, Cicero and Seneca, have often been taken as a statement of tolerant, large-minded humanism. In pricking the grandeur of the sentiment, Beckett's 'The sport of kings is our passion . . . nothing human is foreign to us, once we have digested the racing news' nevertheless maintains intact the contrast between narrow-minded accountancy and benevolent solidarity.[2] Yet, in fact, accountancy provides precisely the original context of the utterance. Chremes is at this point in the play defending himself against accusations of meddling in the affairs of his rich neighbour Menedemus, the 'self-tormentor' of the play's title, who has set himself to unnecessary toil on his estates in remorse for driving away his son with his severity. What Chremes is concerned with in his neighbour is not primarily his happiness or well-being, but the absurdly unprofitable nature of the bargain he is striking with himself:

> You're sixty years of age, or more, at a guess. Nobody in this area has better or more valuable land. You have quite a lot of slaves, but you perform their tasks yourself as assiduously as if you had none. I never go out so early in the morning or return home so late in the evening that I don't see you on your farm digging or ploughing or carrying something. In short you never slacken off or take any thought for yourself. I'm pretty sure you're not enjoying this way of life. 'Well,' you may say, 'I'm not satisfied with the amount of work that's getting done here.' ['at enim' dices 'quantum hic operis fiat paenitet.'] But if you spent the effort which you spend on doing the work on putting them to work, you'd be better off.[3]

So it looks as though, for Terence at least, at this moment, the statement of the most absolute and inclusive human sympathy is focused by questions of quantity and measure.

The UK *Daily Mail* newspaper reported on 6 August 2015 under the headline 'POLICE WILL NOT PROBE BREAK-INS AT HOMES WITH ODD NUMBER' that, in an effort to save resources, Leicestershire police had been sending out teams to investigate reported break-ins only if the house number in question was even. The report quoted Gavin Hales of the Police Foundation, an independent think-tank that studies policing, saying: 'The notion of denying 50 per cent of victims a basic service, based on something as arbitrary as their house number, looks ethically dubious at best.' Meanwhile, Leicester South MP Jon Ashworth called the move 'ridiculous and haphazard'.[4]

Let us assume charitably that Leicester police are not reducing their burglary support in order to free up more time for inter-constabulary football matches, but in order to be able to increase the amount of resource for other, possibly more effective services. So at issue really is not the reduction of services. The outrage attaches to the method used to determine which homes are visited. Commentators were said to be appalled that the service might be reduced through means that are 'arbitrary' or 'haphazard'. But if the service is to be reduced, what possible other basis might there be? Ethnic origin? Debt status? Length of residence? Tax bracket? Criminal record? Day of the week? What is wrong with the scheme is not that it is arbitrary, but that it is not arbitrary enough – since it might in time encourage the belief that people living in odd-numbered houses could henceforth be plundered with impunity. The only way to be sure the reduction was applied evenly would be to alternate between the odd-number rule and the even-number rule at random, perhaps through flipping a coin.

We do not seem to believe that public services, which we experience as particular individuals, should be managed statistically, in terms of large numbers. It is one of the clichés of healthcare reporting in the UK that it is scandalous for there to be a 'postcode lottery' when it comes to things like cancer treatment. But if it were in fact necessary for there to be variation in treatment and outcomes (and, though variation is certainly undesirable, it is just as certainly unavoidable), it would actually be much better for the variation to be the result of a lottery than the result of socio-economic variation, which is really what is being complained about with the phrase 'postcode lottery'. These concerns are a sign of how puzzling but also how pervasive and intimate our relationship is with number in any collectivity in which the gap between large and small numbers is significant. It is often said and no doubt also felt that human beings living in large collectivities are more and more subjected to purely numerical or statistical considerations, but the truth is that human beings, like all organisms in this bit of the universe, have never lived apart from number, but always lived at the intersection of two orders of number, the small and the large. We all want to be treated justly, but justice, at least in the Kantian perspective, involves trying to make small and large numbers accord with each other, not trying to remove numbers from the equation, or to remove any kind of equation from our reasoning and what we feel about our reasoning.

I am going to propose in this book that we have no choice but to make accommodation to number. This will turn out, I think, not to be nearly as bad as it may seem, since it is the case not only that we in fact want no other choice, but that this is a choice that we have already made, for a very long time, and on many different fronts. We could do with understanding better the subtlety and pervasiveness of the agreements that we already make with numbers, in all their dimensions, of quantity, magnitude, frequency and risk.

INSIDE THE AVALANCHE

Ian Hacking refers memorably and quotably at the beginning of his *The Taming of Chance* to the 'avalanche of printed numbers' that became part of the experience of nineteenth-century Britain.[5] Hacking's book addresses itself principally to the history of statistics, and it is as part of this history that number and numbers have usually been considered. It is as a result of the faith in statistics, at least in the early part of the nineteenth century, that it can be said that 'Numbers were a fetish, numbers for their own sake.'[6] This association of numbers with statistical thinking means that they are subject to the same reading as Hacking applies to statistics as a whole, namely that they represent a 'machinery of information and control'.[7] In a world in which laws began to be represented as numerical constants, you could 'collect more numbers, and more regulations will appear'.[8] Hacking's is only a more Foucauldian and uncompromising version of a tendency that is apparent through most writing on what William Petty had called 'political arithmetic'.[9] The taming of chance means that a subject that has its basis in error and variation as much as in norms and regularity is pushed towards control:

> The taming of chance seems irresistible. Let a man propose an antistatistical idea to reflect individuality and to resist the probabilification of the universe; the next generation effortlessly coopts it so that it becomes part of the standard machinery of information and control.[10]

This has been an addictively congenial picture for many who have recognized and attempted to set themselves against the growing dominion of number. But the use of numbers for government is not the whole truth, and, for my purposes, not the most interesting or important thing about the fate of number in the nineteenth century. For, whatever the unease and revulsion it may have caused, Hacking's 'avalanche of numbers' ensured also that numerical consciousness and what might be called quantitative affect became pervasive. If in one sense mathematics secures the division between art and science, in another sense, mathematics increasingly occupies a strange position, as a third thing, reducible neither to science nor art, yet in a sense participating in both. Words cannot be the simple opposite of number, since, let us not forget, numbers are also words. It will become clear in what follows that words, like everything else, are becoming increasingly number-like.

So, far from experiencing an avalanche of number, a simple white-out, number has entered into quality. Quantity has become qualitative. To speak of an avalanche of numbers, as Hacking does, is to evoke some uniform, indeterminate substance, like mud or snow, under which one may be crushed or asphyxiated. In what I have to say, this deathly indifference will be an important part of the experience or apprehension of number; but it is precisely because of this quality of indifference that number is able to become so polymorphously alive, that the powers and qualities of number can be so various. It is precisely because numbers are all the same that they can become so plural and can be the vehicles of so many kinds of value. Without number, there can be no growth, no transformation, no invention, no desire, no value. Yes, numbers can seem to be the dealers of death, even, as Chapter Three, 'Horror of Number', will propose, to be identifiable with death itself: yet without number, life is scarcely conceivable.

There are numbers everywhere in modern life. Numbers are used to identify – service numbers became common during the First World War, and commodities were increasingly identified with serial numbers from the 1920s onwards. Passports, bank accounts, insurance policies, share certificates, even banknotes started to have serial numbers. It is just because of this omnipresence of number that numbers are not only used to put or keep people in their place. Sports fans live a life of impassioned statistics. The discourses of probability and risk provide a shimmering background to every decision that we make. People strive to amass money, but they also have target weights and personal bests. Getting and spending does not always or merely lay waste our powers; it also occupies, exercises and elaborates those powers. The work of calculation is not abstract, inhuman, anonymous and dominative, or it is not merely these things; it is part of the changing texture of our engagements and investments, our dreads, dreams and desires.

Under these circumstances, the argument that the order of number is set over against free, spontaneous existence, acting always as its impediment or cancer, or as a form of social control, that number, in short, is deployed as a mechanism of Power against Life, must seem ludicrously, almost pitiably crude. Nevertheless the idea remains deeply embedded, even systematic in contemporary thought, an idea that we think with too implicitly for us to be able to think about it. Against this idea, I will have to try to show two things. The first is that word and number have, since at least the beginning of the seventeenth century, been drawing ever closer together, and indeed have become more and more entangled, if never exactly to the point of indistinguishability (that would no longer be entanglement). Indeed, the various ways in which number has been projected or excreted as the antagonist of life are the by-products of this commingling. The second is to show that there is – both as a consequence, and in any case – no pure realm of number, that the idea of pure number is in fact a fuzzy and primitive approximation. Language must include mathematics, because it must include the language of mathematics – numbers are signs, and there are no numbers which cannot be signified, along with plenty of numbers that can *only* be signified, because there will never be any way to see them. More controversially, mathematics must include signs, because without signs there is no proof. Words and numbers are not extrinsic to one another.

This means that there is no autonomous realm of number. It follows that there can be no intrinsic being of number either, no one thing that number essentially is, either in a negative or in a positive sense. Both of these bad ideas are energetically ventilated in the work of Alain Badiou, currently among the most influential of contemporary Platonists. According to Badiou, the degraded form of number is the being-in-the-world, or perhaps the simple *being-the-world*, of 'capitalism'. Indeed, what is held to be most horrifying about

Albrecht Dürer, *Melencolia* I, 1514.

capitalism is its alleged 'reduction' of all human and natural possibility to numerability. Against this, Badiou demands and offers us an ontology of number, the claim that number is the truth of being itself. It is very hard to make much sense of this claim, which seems to be sustained largely by the caustic force of Badiou's scorn for all the merely derived or inessential forms which number or being may take, which, as in Plato's theory of absolute forms, seems to make it necessary, on the rebound, that there be ideal, absolute and immaculately self-equalling numerical forms.

Number, like language, is an open totality. They are both systems under construction, which, as they ramify, inevitably become less identical with themselves. Increasingly, not only is nature written, as Galileo wrote, in the language of mathematics, *language* is written in the language of mathematics; and the kind of language we call literature is closer to that kind of mathematical language than other kinds. *Numerus* is not *numen*; not etymologically, and not philosophically either. Latin *numerus* may be an ablaut variant of the same Indo-European base from which Greek *nomos*, law, usage, melody, derives, both cognate with Greek *nemein*, to distribute or manage. Number's name is therefore first cousin to the name of name. The study of numbers might very well be called *numeronomy* rather than *numerology*, since it is the study of usages (and melodies).

In fact, one way to characterize the outlook I want to try to encourage is as an anti-numerology. By numerology, I mean the exercise of number-magic. Numerology is not so much the belief that number governs everything as it is the belief in the non-mathematical powers of certain numbers – lucky numbers, magic numbers, 'numbers of power', the Number of the Beast and so on. It is very hard for anybody interested in number to keep their head clear of the number-magic of numerology. The usefulness of prime numbers in encryption, a topic that has become important for us recently, can encourage magical attitudes towards these numbers, and, indeed, mathematicians themselves are oddly prone to a kind of obsessiveness about certain kinds of number. But numerology can

Detail from Dürer's *Melencolia* I showing a magic number square.

tell us nothing about numbers, even if it can tell us plenty about the power of number over us, a power that we ourselves give it, in the duality that characterizes much magical thinking.

It is very easy to fall into a numerological delirium that strives to see art, literature and culture as governed by or expressive of putative mathematical absolutes – Platonic solids, the Golden Section, that sort of thing. This lamentable tendency is clearly illustrated by Quentin Meillassoux's study of Mallarmé's 'Un coup de dés jamais n'abolira le hasard'. The first part of the study aims to show that Mallarmé's poem is in fact completely governed by number, in a way that also allows it to be understood as a reflection on its own status as poetry, caught between the principle of regular metre (embodied for French poets in the twelve-beat of the Alexandrine line) and the unpredictable patterns of free verse. Meillassoux wants to explain how it is that a particular number could be number itself, which he regards as equivalent (regarding things as equivalent is indispensable to his method) to being identical with chance. 'If we obtain the Number that can be identified with Chance, it would possess the unalterable eternity of contingency itself.'[11] After much hushed lighting of candles and laying out of ceremonial objects on Meillassoux's part, it turns out that the way in which Mallarmé will achieve this is by making the metrical unit of his poem the total number of words in it. It will thus constitute contingency as destiny, presenting itself at once as free and constrained. It turns out that there are 707 words in the poem (there aren't really, so Meillassoux has to fudge the count). Seven is chosen because it 'represents a medium term between the classical metric and pure chance – 7 rhymes in a sonnet, seven stars in the Great Bear or "Septentrion"'.[12] The number 707 is special because it seems to enact a doubling or rhyming with itself around or across the nothingness of an abyss – identified both with the whirlpool of the blank ocean and the central fold of the central opening of which the poem consists. It is therefore 'the Meter by way of which 7 rhymes with itself across the gulf that separates it from itself'.[13] Meillassoux's analysis has little to tell us about number in itself, but is a delightful object lesson in what we allow ourselves to do to and with the idea of number.

If we are to understand the richness and the variety of the ways in which number and quantity work on us, we need first of all to put plain numerality in place of numerology, accepting the principle of number's indifference that is articulated in Margaret Cavendish's *The Blazing World* (1666) by the spirits who respond to the Empress heroine's enquiries about the nature of number:

> Then she inquired, whether there was no mystery in numbers? No other mystery, answered the spirits, but reckoning or counting; for Numbers are only marks of remembrance. But what do you think of the number four, said she, which Cabbalists make such ado withal, and of the number of ten, when they say that ten is all, and that all numbers are virtually comprehended in four? We think, answered they, that Cabbalists have nothing else to do but to trouble their heads with such useless fancies; for naturally there is no such thing as prime or all in numbers; nor is there any other mystery in numbers.[14]

The Empress finds this hard to take, and tries out a few more numerological doctrines:

> Then the Empress asked, whether the number of six was a symbol of matrimony, as being made up of male and female, for two into three is six. If any number can be a symbol of matrimony, answered the spirits, it is not six, but two; if two may be allowed to be a number: for the act of matrimony is made up of two joined in one.[15]

Even when the Empress raises the stakes to the highest level, by asking about the number of God, her interlocutors persist in their patient rebuttals of her desire for numerical mystery:

> She asked again, what they said to the number of seven? whether it was not an emblem of God, because Cabbalists say, that it is neither begotten, nor begets any other number? There can be no emblem of God, answered the spirits; for if

> we do not know what God is, how can we make an emblem of him? Nor is there any number in God, for God is the perfection himself, but numbers are imperfect; and as for the begetting of numbers, it is done by multiplication and addition; but subtraction is as a kind of death to numbers. If there be no mystery in numbers, replied the Empress, then it is in vain to refer to the creation of the world to certain numbers, as Cabbalists do. The only mystery of numbers, answered they, concerning the creation of the world, is that as numbers do multiply, so does the world.[16]

The power of number is an open power, and a power of opening. If this is not a book that will unveil the secret power of certain numbers, then neither is it any kind of contribution to mathematical theory or the philosophy of mathematics. I do however think there is a great deal to interest us in the ways in which people think and write about these topics, especially if they themselves are involved in them, because such thinking is part of the rich mulligatawny of ideas that we have of mathematics, even, we might say, the fantasy that we have about mathematics. And, heaven knows, mathematics is saturated and supercharged with fantasy, not least because it is supposed to be the area of mental life in which fantasy is reduced to its absolute minimum; there is no more self-indulgent idea than the idea of 'rigour' and no more imperious fantasy than the fantasy of escape from fantasy.

In *Where Mathematics Comes From* (2001) George Lakoff and Rafael Nuñez have sought to inaugurate 'the cognitive science of mathematics', arguing that 'Mathematics is deep, fundamental, and essential to the human experience' and that, though 'crying out to be understood', it has not been made sense of in cognitive terms.[17] I want with this book to make a similar move with regard to the cultures of number, opening and extending the understanding of the rich, tough, subtle imagination of quantity, extent and magnitude that runs through human life. I am interested in getting more people more interested in the ordinary ways in which number is lived, which is to say the affective, psychological, oneiric, economic, sexual, ludic and sexual life of number, for which open series we might allow the composite expression, the 'cultural life of number'. Where Lakoff and Nuñez are interested in where mathematics cognitively comes from, my concern is with vernacular mathematics – how we do things with numbers and how they do things with us.

2

QUANTALITY

DOING MATHEMATICS

Some of us study mathematics, but all of us, to some degree, do mathematics all the time. We do not say that we *do* literature or geography or physics, unless we mean that we are studying them at school. But mathematics always involves a kind of doing. It is for just this reason that some more high-minded mathematicians, the kind for whom mathematics is a matter of concepts rather than calculations, tend to see the doing of mathematics as a wearisome and rather vulgar necessity.

On the face of it, the difference between being and doing mathematics ought to be clear. There are countable numbers everywhere, in flocks of birds, tree rings, cycles of the moon, fluctuations of temperature, births, deaths, salaries and marriages, but the enumeration of those numbers is a distinct operation. What we call mathematics occurs almost entirely on the side of enumeration, of counting and the counted rather than the countable. There may be numbers everywhere in nature, but they are not really numbers until they have been numbered, made explicit as numbers or told off as countable things. This depends upon what Alain Badiou has called the operation of the 'count-as-one', the possibility of treating things as though they were equivalent and therefore addable units, and what we might on our own account call counting-as-countable.[1]

Numeration requires an apprehension of number as an abstract system, which allows the 'threeness' of three trees, three sheep and three stones to be recognized. Oddly enough, one might think, this recognition is not itself necessarily numerical, for it depends upon the principle of tallying. This is the system that allows a shepherd to keep count of the members of a flock of sheep without having to count them, by matching the members of the flock one by one to a series of notches on a tally stick, or an umpire to count off the balls in an over with the coins he transfers from pocket to pocket. The shepherd need not have any idea at all of how many sheep he has in his charge, or indeed how many may at any point be missing. He just needs to button up one series to another parallel series. What is more, it should be apparent that this kind of one-to-one matching is in fact what occurs in an act of counting and in any operation in which numbers of things are lined up with each other. Such matching may be as much geometrical as arithmetical, for it may allow or require us simply to see that certain shapes or quantities pair up without remainder. All weights and measures are defined in terms of matchings.

If we want to say that this kind of matching is not really mathematical, because one is not being mathematical unless one is working with numbers, considered as numbers, as opposed to simply matching things, one would also have to admit that none of the machineries we employ to help us do our mathematics, or even do it for us, such as the fingers or toes, the tally stick, the abacus or the computer, are themselves mathematical, in that they all just match things up rather than really discerning or determining relations. They certainly seem to do mathematics but cannot be said to know that they are doing it, and it might seem that knowing what you are doing is definitional when it comes to mathematical reasoning. But then it becomes exceedingly hard to say what there is in calculation that is anything other than these actions of matching, only performed not with shapes but with quantities, in which it is not five fingers or sheep that are of interest, but five fives, of anything. Still, however, the principle of matching persists, in the very fact that we do not have to count every five, just to check groups off.

A great deal of this kind of matching occurs in the natural world – wherever, for example, questions of fit, coupling or concordance may be important. Antibodies do not really recognize foreign organisms, but they do form bonds of correspondence with them, which involve matching of elements. The most important form of matching is surely pairing, of which there are probably two principal kinds. In one kind, distinct things come together to form a couple, or a two that counts as one, as when two animals mate. In the other, some entity duplicates itself, as, for example, when a sequence of genetic code makes copies of itself. These processes, in which two becomes one and one becomes two, themselves come together in sexual reproduction. Such matching is common in aspects of the more formalized arts, like poetry and music, in which we may be required to recognize relationships of correspondence and divergence.

GIVEN TO NUMBER

Can there be numbers in themselves, prior to and separate from the act of counting them? What does it mean to say that there is, or there are, a certain number of crows on a wire, apples in a tub, or atoms in the known universe? Our hesitation over whether to attach a singular or plural verb to the phrase *a certain number* may help us think about this. For we always have the choice to make about whether to think of a number of things as a multiplicity or as a singularity, as a spreading out or a summing up. It is a choice as to what we intend to do with it. To count something is to draw out one aspect of its nature. It is to spell out or make explicit at least one aspect of what it is. What do I mean when I say that I have ten digits, that there are two types of camel or seven colours in the rainbow? Quantitative statements of this kind are all in the subjunctive rather than the indicative: if you affirm that there are ten of something, you are saying that, were you to count them, following the rules of counting (distinguishing items which will count as one, counting everything only once, and telling them off according to some accepted sequence of number words), you would find that when you got to the last digit, you would be uttering the word 'ten'. Furthermore, if you were to count the items again in just the same way, you would get the same result. To say that there are ten digits, two camels or seven colours is actually to lay a wager as to all the things that would happen under these defined circumstances. In this sense, a number is a prediction masquerading as a predication.

So something happens to the object subjected to the act of counting, something that both is and is not quite a statement of what it is already. A parable by Jorge Luis Borges, 'Argumentum Ornithologicum', reflects on this question. It is short enough to be quoted in full.

> I close my eyes and see a flock of birds. The vision lasts a second, or perhaps less; I am not sure how many birds I saw. Was the number of birds definite or indefinite? The problem involves the existence of God. If God exists, the number is definite, because God knows how many birds I saw. If God does not exist, the number is indefinite, because no one can have counted. In this case, I saw fewer than ten birds (let us say) and more than one, but did not see nine, eight, seven, six, five, four, three, or two birds. I saw a number between ten and one, which was not nine, eight, seven, six, five, etc. That integer – not-nine, not-eight, not-seven, not-six, etc. – is inconceivable. *Ergo*, God exists.[2]

Borges's mock proof depends upon the belief that there cannot be indefinite numbers. Numbers, by definition, must be definite, that is, must be some number or other. There is the number three and the number four, but there is no number three-or-four. Borges's cod God fills the gap between the indefiniteness of a certain number that is not yet certain (the certainty that the number of birds must be some number or other that is more than one and less than ten) and the uncertainty as to what particular number that is. But perhaps we do not need this gap to be filled; or, even if we do, it is perhaps a gap that not even God can fill. For perhaps this is the gap in which number happens, the interval in which a certain number can be made over into a certain *number*, by some action of counting, or one that will count as one.

Borges thinks (feigns to think, surely) there cannot be a world in which there is a number that is not a definite number, and so calls upon God to keep the score. More strictly, he calls up a God to perform this function, a God who is ultimately no more than the function of the one who knows, or will have known, the number of everything in advance of its being numbered. Otherwise, things will both be and not be what they are. But what if there is such a world, a world that is in fact simply *the* world, in which things indeed exist in two conditions at once, between the condition of being a certain number that has not yet been subject to enumeration, and the condition of being able to be so numbered? Once something has been numbered, it will henceforth always have been the number that has been counted out. Until that point, however, it is only potentially that number. Perhaps everything participates in this movement in which the potential is made actual, or the implicit is made explicit. God is the name either for what would occupy the place of time, or what could exist unchangingly throughout time. But if one allows for the reality of elapsing time, rather than using God as the bridge between the not yet defined and the definite, then the action of rendering some number or other of something explicit as a particular number is a good example of the process of emergence whereby things that are not yet become what they henceforth will have been.

Number usually participates in this emergence, and may even be part of emergence itself. Rather than being written in the language of number, nature allows, implies, excites and undergoes translation into number. Given the existence of some entity given to numbering, nature is given to number. If everything in nature is temporal, then the direction of time is usually towards number, and through number. Number is that to which, and through which, time moves, for time is nothing without, and so nothing but, the movement of nothing into number. Time is not only necessary for number to emerge, number is equally necessary for time itself to be able to pass, or to be the movement that it is. For time to pass, there must be entities by which one might tell the time, where telling means counting as well as recounting: when Nathaniel Fairfax needed a Germanic word for mathematics in his strange project of delatinizing metaphysics, he called it 'talecraft'.[3] Narration and number seem to be tightly twinned: as James Joyce has it in *Finnegans Wake*, 'haven't I told you, every telling has a taling, and that's the he and she of it.'[4] There must be distinction, distance and difference for there to be time, and number provides the primary language of that distinction. I can only cease to be one thing and become another thing, or be a thing in one state and become that thing in another state, through a process that produces a countable result, or may do, as one becoming two. As

A. N. Whitehead remarks, 'Arithmetic of course enters into your nature, so far as that nature involves a multiplicity of things.'[5]

Numbering is just one example of the many ways in which things can be 'prehended', or taken to be something or other. Human beings have progressively been more and more the principal agents of taking things to be mathematically, though they have never been the only such agents. The things themselves cannot be said to be already what they will become and therefore turn out to have been. That is, they cannot be what they are solely and purely in themselves, because what that is will always turn out to depend on some other entity, operation or set of disclosing conditions acting upon them by taking them to be in some way or other. Mathematical reasoning provides the best example of this necessary and inescapable exteriority. Mathematics is incurred by things rather than inhabiting them.

Most mathematicians are Platonists in that they believe that mathematical truths are given and eternal, which must mean that they are already, somehow, even maybe somewhere, in existence. For such mathematicians, mathematical truths are worked out in the sense that they are driven out from hiding, rather than undergoing some change into themselves by being brought out of latency into actuality. But where are all the places of π, exactly, or all the prime numbers? It is hard to believe they are really stored up somewhere, as though on some celestial, super-cerebral hard drive. Philosophy of mathematics divides between those who believe that things like prime numbers are disclosed by mathematical reasoning, and those who believe they are produced by it. What I have been arguing puts me in the second group. The word 'produced' should not be understood here to mean arbitrarily fabricated out of thin air. The decimal expansion of an irrational number is produced in the sense in which a play is produced – it is drawn out of a script, or a set of prescribed conditions, which limit without fully determining the actualization of that script, which will always nevertheless be the making-actual of that script specifically. Every *Hamlet* is a different *Hamlet*, but, and even because, all of them are stagings of *Hamlet*. So, in a certain sense, the production of a number or a proof or a mathematical result is indeed a disclosure, and a disclosure of what necessarily had to have been the case all along. But the little tuck taken in by that tense – 'what had to have been' – is an indication that what is thought of as the immemorial pastness of the truth that is disclosed is a product of an unfolding history. One may think of this as a kind of slow retrieval of what already lay latent, in which case time is seen as a vast process of curling round backwards on itself. Or one may accept, as I believe we would benefit from accepting, that time is the process through which the a priori is produced a posteriori.

The number of something is therefore not part of what it is, except insofar as what that is is its numerability, mensurability, ponderability and so on. Every present tense is really an elliptical future perfect – and elliptical in two senses. First of all, there is the sense that there is always some ellipsis or elision, something unspecified in what something is. Second, that unsaid or omitted thing may be part of how what is may loop out and come back to itself, as though taking an elliptical detour through the will-have-been. One has one's say about what something is through a kind of shoelace-tying manoeuvre; put your finger on the knot, and hold it artificially in place until you have performed the operation of turning it over on itself, or running it through itself, that will enable it to hold itself together. We create inherence, the way in which things seem to hold together, through making things coherent, matching them up with other things. If I want to understand the nature of war, or love, or intelligence, or a zebra, I put a bookmark in the place where it is, until I have completed a series of operations that will fix it in its place relative to all the other kinds of thing that it might be, so that it then seems to function as its own bookmark. Numbers are examples of Iago's 'many events in the womb of time, which will be delivered', rather than already-existing entities.[6] The thing we call an object is not something that exists already, but something that keeps recurring, as we recur to it:

> An object is an ingredient in the character of some event. In fact the character of an event is nothing but the objects which are ingredient in it and the ways in which those objects make their ingression into the event. Thus the theory of objects is the theory of the comparison of events. Events are only comparable because they body forth permanences.

> We are comparing objects in events whenever we can say, 'There it is again.' Objects are the elements in nature which can 'be again.'[7]

None of this is to say that there is no difference between the numerable and the non-numerable. There are some things which yield much more readily to the possibility of being numbered. But there can be nothing that is entirely immune to being numbered, if only because anything is capable of being seen as one thing rather than many. These conditions are clearly themselves mutable, since they will be dependent upon the capacities either of numerating agents or of numerating conditions to come to light, where by a numerating agent is meant some entity capable of keeping count, and by a numerating condition is meant some set of conditions capable of producing numerate relations. Any field of probability can act as such a set of numerating conditions, making it more or less likely that some outcome or other may come about. This is both thematized in and dramatized by the Parable of the Sower:

> Behold, a sower went forth to sow; And when he sowed, some seeds fell by the way side, and the fowls came and devoured them up: Some fell upon stony places, where they had not much earth: and forthwith they sprung up, because they had no deepness of earth: And when the sun was up, they were scorched; and because they had no root, they withered away. And some fell among thorns; and the thorns sprung up, and choked them: But other fell into good ground, and brought forth fruit, some an hundredfold, some sixtyfold, some thirtyfold. Who hath ears to hear, let him hear. (Mark 4:3–9)

Probability here comes close to probation, trying out, testing or proving. The parable moves from a dim and unformalized intuition that sometimes things work out better than at other times, through to a more and more precise array of the possibilities, in order of desirability – no germination at all, quick germination without growth, germination followed by sustained growth and assured multiplication. The parable then suddenly doubles over on itself, to offer hearing and understanding the parable as an exact parallel to what befalls the seeds in it. Asked why he speaks to them in parables, Jesus explains that it is a kind of distribution mechanism, designed to pick out those who not only hear but understand what they hear. It is usual to presume that the exchange originally written in Greek and here rendered in English would have taken place in Aramaic, but if so it is as though Jesus were aware in advance that the Greek word *parable* itself implies a certain kind of casting or throwing (παρα- alongside, + βολή, casting, throwing). In both the literal and the metaphorical fields there is an automatic counting, or exterior computation, in which what is not known comes to be known by being numerically distinguished, in, or by, some landscape of likelihood.

Perhaps the very word 'field' embodies this history. If a field means any open space, it also suggests some 'opened space', some space of defined openness, a space thereby pushed towards the condition of equiprobability. The word *field* may be etymologically cognate with Greek πλατύς, broad, and Latin *planus*, flat. A field is a demarcated openness; it is a space in which certain variations are drastically limited in order that other variations may be augmented. A field is in itself a computational machinery, perhaps even the kind of machine of white or maximally multiplied possibility that a white page (Latin *pagus*, field) or a blank screen can be.

QUALITY OF QUANTITY

So number is a liability or a tendency, not a fixed and final condition. We live in an era in which a series of linked and reciprocally reinforcing developments, theoretical and practical, have accelerated the process of making the world more and more quantifiable, and more and more a field of numerical operations. Theoretically, forms of mathematics have been developed, from probability to calculus to fluid dynamics and beyond, which enable mathematical account to be taken of natural processes. Practically, a vast array of instruments has been developed which allow for quantitative accounts of processes, including psychological processes, to be developed, augmented by computing machineries that are able to calculate ratios and relations far faster and more accurately than we can. There are many who view such developments with a kind of angry panic, and do everything they can to define and defend the shrinking realm of the nonnumerable from the action of numbering. We need not here be delayed by the question of the truth, or adequacy, of the claim that there are qualitative truths that are under no circumstances renderable as quantities, pleasant though it would be to dally. What we will have to do with will be rather with the experience of having to see, or, what may be the same thing, being increasingly able to see, the natural and human world under the aspect of number, the experience of the dwindling (a dwindling that might itself not be beyond quantification) of the set of things that it is any more in principle impossible to count, or render as number. My concern is with the kinds of adjustment that we, by which I mean primarily non-mathematical persons, are having to make to this world of measurable quantities and calculable ratios. For this reason, my aim is not to quarantine quality from quantity, but rather to articulate some of the specific qualities of the quantitative, some of the many and changeable ways in which quantity comes to exercise its purchase upon us, and we our prehensions of quantity. For this, I propose the term *quantality*, with, as though that were not yet bad enough, *quantical* as its adjectival complement, to imply the tendency or aspiration to render things in terms of quantity. A quantitative analysis deals with known or knowable quantities; a quantical attitude seeks to make out the values of quantity as such. The quantical might be regarded as the subjunctive mood of the quantitative, this slight unsteadiness of meaning assisted by its evocations of words like *nautical* and *quizzical*. Quantality is the name for the apprehension of quantity in general, before (and after) specific quantifications. Quantality is the quality of the quantitative.

In physics, quantality refers to the quantum view of matter opened up by Max Planck, that energy and matter at the smallest scales are not continuous, but must jump between discrete energy states. Even in physics, though, the word is something of an exotic. In fact I first found it used by Oliver Sacks in his book *A Leg to Stand On* (1984) to describe a distinctly exotic experience, that of having to incorporate back into his body schema a leg that he had injured badly while fleeing from a bull on a Norwegian mountain. In the chapter entitled 'Solviter Ambulando', Sacks describes how, following a long convalescence, he had to try to start walking again. The experience he evokes is rendered as the capture of the moment between the formless and the sense of relative measure that is implicit in inhabiting a body, or feeling some body part to be indeed part of one's body. Sacks experiences a chaos of sensations, but carefully insists that this is not raw sensation so much as a kind of uncontrolled delirium of estimations, or quasi-calculations, the feeling not so much of performing calculations as of calculating going on, as he tries to get the measure of a limb that does not yet fit into the world or his body:

> As soon as the tumult of sensations and apparitions burst forth, I had the sense of an explosion, of an absolute wildness and chaos, something utterly random and anarchic at work. But what could produce such an explosion in my mind? Could it be a mere sensory explosion from the leg, as it was forced to bear weight, and stand, and function, for the first time? Surely the perceptions were too complex for this. They had the quality of constructs, and not of raw sensations or sense-data. They had the quality of hypotheses, of space itself, of those elemental a priori intuitions without which no perception, or construction of the world, would be possible. The chaos was not of perception itself, but of space, or measure, which precedes perception. I felt that I was bearing witness, even as I was undergoing it, to the very foundations of measure, of mensuration, of a world.[8]

The miraculous quality for Sacks has something to do with the eruption into consciousness of a process that had long ago been internalized, as a result of all the many complex projections and adjustments we must all have made as we learned to walk for the first time:

> And this perception, or pre-perception or intuition, had nothing whatever to do with me – it was proceeding in its own extraordinary and implacable way; which started, and

> remained, essentially random, while being modulated by some sort of matching or testing, a targeting or guessing, perhaps a trial-and-error process, a wonderful yet somewhat mechanical sort of computation. I was present, it is true, but only as an observer – a mere spectator at a primordial event, or 'Big Bang,' which was the start of inner space, the microcosm, in me. I was not actively, but passively, undergoing these changes, and as such could bear witness to what it was like to be present at the founding of a world. A true miracle was being enacted before me, within me. Out of nothingness, out of chaos, measure was being made. The jumping fluttering metrics were converging towards some average – a protoscale. I felt terror, but also awe and exhilaration of spirit. Within me there seemed to be the working of a cosmic mathematics, the establishment of an impersonal microcosmic order.[9]

The realm of bodily measure seems to be both before the beginning of conscious experience and just after that beginning. It makes briefly and incandescently apprehensible, as a sort of corporeal calculus, an experience that lies buried within every moment, and every movement. Sacks looks to quantum theory for the word to describe it – *quantality*:

> All at once I thought of God's questions to Job: 'Where wast thou when I laid the foundations of the earth? Who hath laid the measures thereof?' And I thought, with awe, I am there, I have seen it. The frames, the fluttering frames, made me think of Planck and Einstein, and how quantality and relativity may stem from one birth. I felt I was experiencing the 'pre-Planck time' of myself – that unimaginable time cosmologists speak of – in the first 10–45 seconds after the 'Big Bang' – when space is still unstable, fluttering, quantal: that time of preparation which precedes the beginning of real time.[10]

Learning to walk again, using the Mendelssohn Violin Concerto to tone and tune his gait, helped Sacks grasp the way in which complex dynamic unities both forget and remember the computations of which they are composed: 'Here, in doing, one achieved certainty with one swoop, by a grace which bypassed the most complex mathematics, or, perhaps embedded, and then transcended them.'[11]

Sacks's experience was a kind of visionary explicitation of the computational nature of experience, achieved, not through conscious calculation, but through a kind of immediately self-sensing coenaesthesia. But we are more than ever aware of the computed ingredients of more and more of our apparently undecomposable experiences. There are many more profitable things that the humanities might be doing than what currently absorbs their resources, on such a prodigious scale, despite their claims to have to live like church mice: but one of them might be to take the measure of such movements between implicit and explicit quantality.

But we have in part to deal with the glum paradox that, as more things become quantifiable and in an ever greater variety of ways, so attitudes towards quantity in many quarters harden and simplify. The more quantity might offer to lead us away from dogma, the more calcified the dogmas become about what is called quantity as such. The only quality that many of us are equipped or prepared to recognize in quantification is that of reductiveness (we will not linger on the fact that the idea of reductiveness can scarcely be regarded as any kind of escape from quantitative thinking). It seems obvious to many of us that, if you count something, you reduce its complexity to one dimension alone, taking account only of its numerical aspect, with everything else dropping out of consideration. But this reduction is by no means the end of the process. For the reduction effected by quantification gives access to a vast multiplicity of different kinds of mathematical relation (that, for example, of multiplication itself), performable on different scales and across different periodicities. One might imagine the objection that these relations are nevertheless purely mathematical relations, and therefore lack the richness and complexity of qualitative relations, between things like colours, hopes and difficulties. But ours is a world in which the interchange between quantities and qualitative states is richer than ever before. We do not any more have to regard numbering as final or definitive, a putting-to-death through exactness. Number is no longer the end of any story. To say that we have become more quantitative than ever before is not to say that everything must be rendered up as number, without remainder, and then abandoned: it is to say that number is always in the middle of things. We have been inhabiting such a world, in which qualities and quantities incessantly alternate with and give rise to each other, for some time now, though we do not seem to be as good as we might be at understanding it or its possibilities.

DIGITAL

Of course, there can be no question that number is reductive. When one converts something into numerical form, one imposes homogeneity on to heterogeneity. A digital encoding reduces what it encodes to a sequence of alternating binaries, the simplest numeric system that one can imagine. But having once established such a principle of equivalence, it then becomes possible to produce variations of a much greater complexity than would be possible in any other way. Michel Serres makes a similar observation about the measurement of weather systems. There are, he reminds us, really only a small number of variables of which one needs to take account in understanding and predicting weather patterns: pressure, temperature and wind speed chief among them. But variations in this small number of parameters are all it takes to produce the almost unfathomable complexity of the weather, of which we continue to have only limited, local and short-range understanding.[12] Reduction to number need not mean reductiveness as such. Translating things into numbers or giving them number-like qualities and relations, which essentially means dividing wholes up into smaller parts, hugely multiplies the possibilities of what may then be done with those wholes.

The digital revolution has amplified and itself been amplified by a particulate awareness, an awareness that solid states of every kind are in fact made up of molecular masses. This is not a new apprehension. The earliest atomists Democritus and Leucippus postulated that the atoms of which matter is composed were of every possible size and shape, and the differences between these sizes and shapes were what accounted for the different qualities, of sharpness, redness and so forth, of different kinds of matter. Lucretius gives a memorable statement of this in *De rerum naturae*: 'You may readily infer that such substances as titillate the senses agreeably are composed of smooth round atoms. Those that seem bitter and harsh are more tightly compacted of hooked particles and accordingly tear their way into our senses and rend our bodies by their inroads.'[13] But Epicurus, whom Lucretius follows in many respects, disagreed with Leucippus and Democritus, in arguing that atoms were in fact identical and without qualities, meaning that the qualities apparent on larger scales emerge only as a result of the arrangements and movements of these constituent atoms rather than their intrinsic qualities.

But this implies that the difference between quantity and quality in fact reduces to a difference between scales, the emergence of quality being the effect of a movement from one scale to another. Since scale is itself a measure of quantity rather than of quality, quantity intersects with and enters into quality. Quality is just quantity at low resolution. Technologies like half-tone printing and the cinema often bring the thresholds between perceptual scales into visibility, through the pointillist shimmer and flicker that lets us see the constituent 'atoms' of our vision.

The digital revolution is really only a prodigious acceleration of the many forms of digitization that have been developed through history, if by digitization we mean the decomposition of wholes into equivalent and so exchangeable components. Bricklaying, mosaic tiling and typesetting are all examples of digitization. It has made clearer than ever the ways in which the decomposition of wholes into orderable multiplicities makes it possible to create much richer and more complex variations and relations. If one imagines a simple rod made of solid and impermeable matter, it would be impossible to convert it into any other shape. Break the rod up into a number of joints and it becomes possible to bend it into different shapes. The more divisions one can make in the rod, the more flexible it will be. What is more, the more exactly equivalent these divisions are, the greater the possibility of variability in the elasticity of the rod.

But a further principle needs to be added to that of equivalence for the rod to become maximally transformable: that of ordinality. If the equivalent units can be ordered in a series, whether that series be generated by the principle of adding one unit at a time, as in counting, or some other, more arbitrary principle of ordering, like the alphabetic series, or even the sequence of colours in the rainbow (which may be expressed mathematically in terms of electromagnetic frequencies), it becomes possible not just to order, but rapidly to reorder the series, and not just by counting through, but by the combination of elements. One may think of the rod as a simple form which is made more complex through being divided into permutable units. Mathematizing the rod subjects its simple and absolute givenness to an arbitrary and external principle – no matter how small the units, their size will have to be given by means of some principle that is heterogeneous to the rod.

But one might also think of the rod not as something simple but as a kind of massive chord with all its constituent elements sounding simultaneously, but undecomposably, containing the sum of all its possible divisions and relations. Dividing the rod up into those elements means being able to play these simultaneous possibilities out in time. But it also means having all the elements available as a kind of space to be traversed. The unquantified rod is like all the notes of the piano, and all the conceivable notes between them, being sounded at once. The divided rod is like the keyboard, whose potential internal relations can be actualized in temporal patterns.

Numerical relations are relations between equivalent, exchangeable units, whether or not these consist of numbers: so in this sense a film is already digital, because the separate frames have a numerical relation of absolute equivalence to each other. But numerical relations are not just exactly equivalent, but also orderable and therefore manipulable, because ordering otherwise equivalent units puts them into a navigable space. Numbers are atoms with names.

There are few more powerful enactments of this principle than printing. Printing literally transformed individual characters into separate blocks. Because these blocks were interchangeable, they could be varied, endlessly set, broken up, cast off and redistributed. None of the gains in reproducibility that printing offers are possible unless the type is cast so that the characters are of an equivalent size, which can sit next to each other in the compositor's stick. This converts the analogue forms of letters, based on the principle of continuous variation, into digital forms, made up of patterns of discontinuous units. But, if printing gave this principle material form, it was only able to do it because people were already users of writing, which involves a digitization of speech. Writing broke up continuous events into discontinuous objects; indeed, writing made it possible for speech to be considered as consisting of distinguishable or comparable forms of event, such as ejaculations or sentences. Writing also involved deciding, for example, that the flow of speech could be chunked into units called words, and that those words were themselves configurations of a small number of elements that might be called characters or letters.

But even this digitizing decision is one which every oral language or spoken system of communication seems already implicitly to have taken with regard to itself. For every system of spoken language requires shared and therefore exchangeable forms. This implies in its turn the divisibility of speech into recognizable and repeatable units – that are recognizable because they are repeatable. If it were possible to use a different kind of sound to signify 'mama' every time that concept needed to be signified, babies might do this; but they have to learn that only repeatable speech-events, which is to say digital or numeric-type speech objects or units, will work. There are infinitely many sonic variations in language that might be deployed to signify different meanings, but any language will make certain of these differences salient and operative. This is made clear by the phonemic variations between languages: in English the difference between /l/ and /r/ is semantically significant, whereas in Japanese this is not a phonetic difference that makes any semantic difference. For the units of spoken language to be meaningful it is a huge advantage for them to be equivalent and therefore exchangeable; in a large number of English words, the sound /l/ can readily be substituted for the sound /r/. Phonemic differences are not just differences, they are equivalent *kinds* of differences.

This implies that the digitizing of language is not, as our conventional contrast between the analogue and the digital might imply, the move from one principle to another, absolutely heterogeneous one. Digital encoding is as powerful as it is because it is the extension, or, as we might say, the inward intensification, of a principle of organization that is everywhere at work already in all organized language, if not always to the same degree. And this form of organization is numeric, sharing with number the twin principles of divisibility and orderability, which is to say compressibility. Divisibility tends towards and depends on the principle of equivalence, which makes individual units maximally interchangeable with one another. Orderability makes variability possible, because what can be ordered can be reordered, and to reorder is always easier than to order the unordered: you cannot engineer water, but you can engineer the compound molecules of which water consists. We can say that these numeric systems allow for *variable equivalence*.

There might be other ways to produce the huge variety of forms that physical matter can take, but it is unlikely that any arrangement that did not involve the variable relations between equivalent units would ever produce this variety of property and quality. There might be another way for living creatures to have evolved the vast range of form and function that makes up the design space of evolution, but there seem good reasons why the permutation of just four elements in DNA coding is both necessary and sufficient for this.

The point about reducing something to number is that one is thereby enabled to go beyond simple numbering into the prodigious enlargement of relations that mathematics allows (and let us be assured that it is only from its putative outside that mathematics seems like only one kind of thing, math in the singular rather than plural maths). A. N. Whitehead indicates something like this in his *Religion in the Making*, in defining a dogma as 'the precise enunciation of a general truth, divested so far as possible from particular exemplification'.[14] The example he gives is the notion of irrational number, which had formed part of mathematics, albeit in rather a shadowy way, until it was given accurate definition in the last quarter of the nineteenth century. It is possible to see such formalization as the delimiting of possibility, but Whitehead argues, and surely rightly, that 'such precise expression is in the long run a condition for vivid realization, for effectiveness, for apprehension of width of scope, and for survival.'[15]

QUITS

Numbering has the reputation of making things more precise, thereby taking away the possibility of uncertainty or mystery. But we are wrong to see number always as the exaction of exactness. The word *precise* comes from the past participle of Latin *praecidere*, meaning to cut off: until the seventeenth century, to *precide* meant to excommunicate. In its early history, precision nearly always has this sense of the abrupt or reduced, not to say, on occasion, the homicidal. But it was also commonly used by Protestant theologians suspicious of overly scrupulous forms of religious observance, for whom being 'superstitiously precise' could be a kind of idolatry.[16] Indeed, it is for this reason that the precise could come to constitute a kind of exhibitionist exorbitance, as in Biron's apology in *Love's Labour's Lost*:

> Taffeta phrases, silken terms precise,
> Three-piled hyperboles, spruce affectation,
> Figures pedantical: these summer flies
> Have blown me full of maggot ostentation.[17]

It is a mistake to imagine that precision will always involve the reduction of the complex to the simple, though this may be an important accessory principle. In fact, since the development of statistical understandings of the process of measurement, which emphasize the need to acknowledge the likelihood of variation, precision has come to veer over into its opposite, meaning the propensity of any set of measurements to produce variant results, that is, the amount of *imprecision* that may be expected of any observation or measurement. The first use of the word 'imprecise' recorded by the OED is as late as 1805, though a couple of examples may be found a little earlier in an eighteenth-century book on spelling reform.[18] Francis Thompson complains in a 1907 *Athenaeum* review of Henry James's *The American Scene* of a Bironian excess of precision in the work of James: 'He must still write about and around it, and every way but *of* it – must approach it by stealth and tortuous indirectness, and deck it with the most elaborated precisions of impreciseness, as if it required hinting afar off.'[19] Indeed, Thompson seems to make the Shakespearian connection explicit, groaning that 'throughout four hundred and sixty-five broad pages there is no oasis in the level, unbroken expanse of Jacobean style.'[20]

The particular topic that Thompson instances as requiring the 'accumulation of every Jacobean resource for uttering the unutterable' is that of the effect of modern mass existence: 'He tells you that, among the vast numbers newly cast into the machine (so to speak), the most striking feature is their featurelessness, the dead blank of monotonous uniformity which has resulted.'[21] Thompson's complaint is essentially a statistical one – he would like James to state the general truth straightforwardly and without the mincing application of the 'microtome', a blade used to cut slices of matter for microscope slides.[22] Thompson wants James to go straight to the rough-and-ready average, when the question of how that is derived is the one that seems most absorbing. However taxing the result, the precision of impreciseness seems, excuse me, *precisely* the interesting point about James's response to the phenomenon of mass uniformity.

So, for example, one might say that the principle of air conditioning depends upon the powerful but vague apprehension that human beings, as creatures who have to maintain a constant body temperature, find large variations of temperature stressful and unpleasant, and tend to operate most successfully in a smaller rather than a larger range of temperatures. This might lead one to the specification of an average temperature that would conduce to thermal comfort. However, this kind of precision only discloses the need for further precision, since it becomes clear that variation in temperature is also an important feature in human well-being, in which the principle of 'thermal delight', in Lisa Heschong's superb phrase, is paramount.[23] Precision, in other words, may pass through definition, or the reduction of variation, but does so in order to be able to be more precise about variation itself. Precision, which begins by being categorical, is thereby able to become circumstantial. This is the principal reason why one cannot simply set variation against precision, since variation enjoins precision, and precision discloses variation much more vividly and usefully. Fuzziness is not the enemy of precision; it is its intensification.

This principle may be seen from reflection on the exquisite adverb *quite*. This is one of those primary words which interested Freud, in which the very force and reach of a word's application mean that it slowly spreads across into its opposite.[24] Thus *quite*, which means completely or entirely, is equally likely to mean almost or only to a certain degree. When used to signal approximation, *quite* actually means *not quite*. The difference, appropriately, is detectable in English only by a slight phonetic modification of stress, in the difference between being quite *certain* and *quite* certain (this is a quadratic equation, of course, since the word *certain* is subject to the same fluctuation). The word *quite* seems to be related to the word *quit*, which signals the condition of being free from a debt or obligation, the force it has in words like *acquit* or *requite*. This might tell us that the primary meaning of *quite* is actually explicable as a sort of duality, as that which has been cleared or equalized, or reduced from something to nothing by being paid back or discounted. *Quite*, in fact, requires a notion of equity, or the undoing of an iniquity: there is always an implied doubling in being quits, so that, as Phebe notices in *As You Like It*, 'omittance is no quittance.'[25] Hamlet's *quietus* brings together Latin *aequus*, equal and *quies*, repose, for, in the legal phrase *quietus est*, one is made quiet by having one's obligations discharged or requited.

This history seems to scurry in the opening words of Beckett's *Malone Dies*: 'I shall soon be quite dead at last in spite of all,' translating 'Je serai quand même bientôt tout à fait mort enfin' in his French original, or perhaps not quite translating it, since the 'quite dead' seems not quite as definitive as the briskly done-and-dusted 'tout à fait'.[26] Garrett Stewart sees in the 'mortally mincing "quite" a preposterous adverbial modification' which carries Beckett's 'sardonic nostalgia for the death scene itself as closural satisfaction'.[27] In a sense, the English goes further than the French, precisely by not going quite so far as to say 'tout à fait mort'. The absurdity of specifying that one might need to be accounted quite *dead* lies in its allowing for the idea that one might otherwise only be *quite* dead, like being a touch pregnant, or somewhat unborn (even though one can in fact be said to be dead on the whole, and perhaps, remembering all those microbes, can only ever be). The quasi-quantitative question of the equivalence of translation, or, at least, its adequacy (the adequate being that which moves towards, but does not quite reach equivalence), here rhymes with the difficulty of knowing when or whether one is fully dead. The English translation has to know that its original has already been laid to rest, even as the very existence of a translation testifies to a grimly disquieting resurrection. If a translation has always to count its original as one, rather than as a process of accounting or recounting that might still be going on, it must always also make one wonder if complete death, or the death of completion, is in fact to be counted on, or can ever constitute enough to be counted as 'one'.

The fact that quietness and quitting are proximal in the word *quite* means that it is often brought into question in the writing of death, since writing both is and is not quite death itself, being both the death and the preservation of living speech. Shakespeare's Sonnet 126 evokes the 'lovely boy' who both grows and yet is held back by a Nature seemingly intent on the 'disgrace' of time. But like the poem, which, in explicating that borrowing of time or holding back from time must also imitate it, Nature must in the end render up its account:

> She may detain, but not still keep, her treasure!
> Her audit, though delayed, answered must be,
> And her quietus is to render thee.[28]

The double-entry of the final couplet not only balances debt with the quittance of death, but makes them equivalent to the question-and-answer duet of audibility and making quiet. The word *quiet* has sometimes provoked reflection on the particular kind of hushing sound it makes, especially in the suggestion that *quietus* might sound like the last breath, as in Keats's invitation to Death to 'take into the air my quiet breath', a line cruelly recalled as Belacqua watches a lobster about to be plunged into boiling water in Beckett's 'Dante and the Lobster'.[29] Emily Dickinson also seems to imbricate quietness with the mortal quietness of that which continues on paper to be able to simulate the airy not-quite-nothing of life:

I breathed enough to take the Trick –
And now, removed from Air –
I simulate the Breath, so well –
That One, to be quite sure –

The Lungs are stirless – must descend
Among the Cunning Cells –
And touch the Pantomime – Himself,
How numb, the Bellows feels![30]

The quality of all these qualifications that the equivocal *quite* permits and prompts is nothing if it is not quantical, which is to say, not quite quantitative, yet certainly not quite not either.

SPELLING OUT

Peter Sloterdijk has argued that modernity, by which he might perhaps mean Dasein itself, insofar as it conceives and thereby produces itself as historical, is itself made intelligible through a process that he calls *explizieren*, explicitation. Indeed, we may see Sloterdijk's own distinguishing of the process of explicitation as an example of the process: the concept of explicitation is an explicitation of its own function. Explicitation is what might be called 'spelling things out', or showing the workings in detail of something.[31] Perhaps we might sum up the process of making the latent manifest with the word *intelligence* – which signifies both the capacity to understand and the process of communicating that understanding. Intelligence, as a kind of making known, will always involve telling, in two senses: the counting out involved in Fairfax's talecraft, and the articulation of that counting. If *explizieren* is the move from inherence to intelligence, it is a version of the view, regularly articulated by Michel Serres, that nature moves from 'hard' form to 'soft' information.[32] The implicit can never be made explicit except through greater precision, and that precision must almost always involve the move from quality to quantity. 'I am almosting it,' thinks Joyce's Stephen Dedalus to himself.[33] Is that quality or quantity?

Sloterdijk sees explicitation as always going in two directions at once. As unfolding, it is part of the 'exodus of humans into the open',[34] the movement outwards from the primal bubbles of co-belonging, of actual and imaginary enclosure, taking as their model the baby in the womb, in which human individuals and collectivities house themselves. It is a movement into number because number is one of the forces that rupture this primal bubble:

> The biune world had known neither number nor resistance, for even the mere awareness that there were other things, countable and third options, would have corrupted the initial homeostasis. The expulsion from paradise means the fall from the blissful inability to count. In the dyad, the united two even have the power to deny their twoness in unison: in their breathed retreat they form an alliance against numbers and interstices. *Secundum, tertium, quartum, quintum – non dantur.*[35]

But number as exteriority is folded back, Sloterdijk maintains, into a second, immunological skin, to contain its very exilic threat and protect against the 'cosmic frost'.[36] One form of this immune system, patching up the ontological rent produced by number, shaping a shelter from the very landscape of one's exposure, is mathematics.

In the great, and late, unfolding of mathematics that has taken place over the last millennium, and especially in the last 400 years, the realm of number seemed gradually to be making itself autonomous of the realm of signs, written and visual. But the apogee of this autonomy, the point at which the orbit of mathematics took it furthest away from the earth, the moment of what Jeremy Gray has called the 'modernist transformation of mathematics', was also the point at which number was about to begin its great re-entry into earthly (that is, social, psychological, emotional, financial, technical) life.[37] As A. N. Whitehead remarks in his *Science and the Modern World*, 'nothing is more impressive than the fact that as mathematics withdrew increasingly into the upper regions of ever greater extremes of abstract thought, it returned back to earth with a corresponding growth of importance for the analysis of concrete fact.'[38] From this position, the lifeworld was infected and inflected by number, even as the very notion of the lifeworld, as a fragile and separated enclave, was more and more an effect of this panicky detraction from number, or the effort to quarantine number from life experience. Though pure or formal mathematics will doubtless continue to become even more forbidding in its complexity and, for that reason, seemingly ever more forbidden to most people, our lives (including, of course, the lives of mathematicians when they are waiting for buses, planting gardens, contributing to pensions, waiting for aeroplanes or fleeing airstrikes) have in fact become ever more densely impregnated by number, and not just in the way suggested contemptuously by Alain Badiou, who believes we are subject to the debased and debasing forms of 'number's despotism', in shopping, elections and cup finals, in which 'what counts – in the sense of what is valued – is that which is counted.'[39] In recasting social and mental life, number has broken free from mathematical reasoning, not least in being automated through digital and mechanical means. Computing technology has not so much alienated us from number as emancipated number from the operations of mind, allowing it to enter more generously and generatively than ever before into different sorts of experience. This does not mean that our lives are governed by number in any simple or asymmetrical sense, but rather that numbers and signs are becoming reciprocally formative and illuminating. I have said that number is the direction in which nature moves – but in fact that movement is not all in the same direction. The move from matter to number, from form to information, is not a steadily rolling river, but tidal, nebular, vortical, polyvectorial.

The most striking manifestation of the compounding undergone and effected by number is the explosion of code, which stands between the conditions of word and number, giving number meaning and performative power, and giving signs (not just verbal signs – think of a barcode) the powers of number. Number no longer lies at the beginning of things, as the Pythagorean primordial principle of the universe, nor at the end of a process in which the numerical laws of nature will have been fully explicated. Number lies in the middle, and has itself grown to be the most energetic mediator of all. We move in and out of number and thereby move number into and back out of explicitness. Before digital encoding – or, to be more precise, before automated digital encoding – number was usually an unfolding of the enfolded, in the specification of the hitherto unregistered, even unsuspected quantities, dimensions, weights, distances, magnitudes, rates and ratios of things. But digital encoding can now enfold as well as unfold, implying prodigious numbers of numerical relations that are no more capable of being registered directly by consciousness than the individual vibrations of musical notes. This kind of encoding does not lead to number as its end point, but employs number as the mediation between sign and sign. We are familiar enough with transformations of numbers into words, as when we explicate what some piece of mathematical notation means, or of words into numbers, as when a set of variant conditions are summarized in some mathematical formula. But we also use numbers to multiply and accelerate the relations between verbal signs, as when I respond to an email, or speak to somebody on the telephone; and can also, though perhaps less often, use words or other verbal signs to multiply or accelerate the relations between numbers. Numbers and signs both translate and potentiate each other. To translate a sign or sequence of signs into digital form is not to fix it, but to mobilize or virtualize it, to allow for translations and transformations that are otherwise unlikely, expensive or impossible.

My focus in this book is on the raw idea of numbers rather than mathematics. Following Christopher Small, musical theorists of an anthropological kind like to use the word 'musicking' to refer to the social practices and contexts of music-making in general, as opposed to the specialized skills of musical composition and performance.[40] The horizon of social quantality within which I propose to move might suggest an equivalent term like 'numerizing' to evoke the inexpert but raggedly resourceful variety of habits, actions and sentiments associated with the workings of number.

How can the humanities, as distinct from humans, benefit from this perspective? Perhaps they are not going to be able to and, if not, then so much the worse for the humanities. For, whatever the humanities elect or neglect to do, humans are going to continue to enter more variously and energetically than ever into number. But if the humanities are to get any benefit, a prerequisite is that they have to find a way to be interested in the topic, which would mean giving up the hysterical institutional allergy to number and all the inhuman, morbid powers it is held to embody. We would have to substitute curiosity for complacent phobia, and would have to want to be intelligent about the particular kinds of intelligibility that number conducts. As the name might suggest (though rather too presumptuously for my taste), the humanities have something to do with the human. But we may take a chilly sort of cheer from the curiously and comfortably plural–singular name of 'the humanities', hoping that, in the uncertainty of its number, the humanities may start to resemble the thing, or the many things, known as mathematics. Many of those in the humanities are appalled by the incursions of the quantitative sciences into more and more areas of human life, principally through economics; indeed, many of those in the humanities take their vocation and make a profession from this denunciation of the quantitative. But the reason that the quantitative sciences have extended their reach is that they have multiplied their forms and their ways of having significance. The mathematics from which so many humanists spiritually and professionally recoil (except perhaps when adding up their expense claims) is a paltry thing compared with the kinds of mathematics that are beginning to be deployed in so many areas of life, many of them areas that one would think would be of importance for the humanities. This is a mathematics that is not just declarative, as Michel Serres has put it, but procedural. Rather than being concerned with the production of general models or formulae, or closing accounts, mathematics is becoming ever more implicated in the operation of functions or procedures. Declarative knowledge is customarily defined as knowledge *that* (knowledge that there are twelve inches in a foot, that Paris is the capital of France, and so on), while procedural knowledge is the knowledge of *how* to do something, such as ride a bicycle. Serres sets out the difference:

> The declarative or conceptual invents ideas, defining them distinctly, follows the principle of reason in its fixed form, following out causes and effects in detail. The algorithmic or procedural constructs events and singularities step by step, entering into details, in a series of times and circumstances.

> The declarative demonstrates, abstractly. The procedural relates, individually.[41]

The difference between the declarative and the procedural is a quantitative difference between the different speeds at which a volume of data may be traversed:

> The algorithmic mode is made possible primarily by computing and other automated procedures, which mean that 'the old image of light changes from clarity to speed' . . . To comprehend thousands of examples, we have less need of the concept . . . Inscribed in the machine, a thousand algorithmic procedures allow the construction and direct envisaging in rich detail of singularities which are no longer smoothed out.[42]

This book is not directly about mathematics. It is most certainly not a philosophy of mathematics. Rather, it is concerned with what happens in the course of this process in which things are brought into the condition of number, with the oscillations between the being and the doing of number, the oscillation between being and doing what number is, the oscillation between being and doing *that* number is. It looks to a condition in which the arts, sciences and other areas of human existence are beginning to converge in a single, though hugely diverse, operation of practical judgement in which all the different forms of calculation cohere, and in which what used to be known so uselessly as interpretation has acquired the new, fair name of engineering.

Not only are the humanities certain to be bound up in number more variously than ever before, if they are to continue to be part of serious intellectual endeavour and not to dwindle into the condition of an indulgently licensed cult, like aromatherapy on the National Health; a steady and serious engagement with number will disclose that the important questions in art and experience have in fact always vibrated with quantical questions: that is, with magnitude, scale, dimension, extent, frequency and duration, along with the acts of measurement whereby we adjust ourselves to them. If there is anything that is truly unextended, it takes its measure always from the *res extensa*. The concern with the nature of personal and cultural identity which has been such an obsession across the humanities is, in principle and action, a reflection on what William James designates as 'the abstract numerical principle of identity, this "Number One" within me, for which, according to proverbial philosophy, I am supposed to keep so constant a "lookout"'.[43] The bad reputation that number has gained has to do with the fact that we are so invested in keeping ourselves whole and entire (that is the mathematical meaning of the word *health*). Human fiendishness seems drawn to mathematics, as in ordeals like being hanged, drawn and quartered, in which the horror of being turned from a complete thing into a series of fractions (as we do to meat) is indulged. The particular cruelty of this punishment consists in the almost comic incongruence between its exact apportioning and its absurd excessiveness, as well as with its duodecimal compounding of multiplication (or division) by three (hanging, drawing, quartering) and by four. To reduce a living body to matter by this method is a sign of the unnaturalness of number, but the quality of the unnaturalness, its mixture of horror, relish and mock-coldness, can be nothing other than human, for only a creature saturated in number could be so attuned to its quantitative qualities.

We need only look at the proliferation of particles like *poly-* and *multi-* in cultural criticism, or the Deleuzian thematics of the molar and the molecular, or the recent prestige gathered by the idea of the 'multitude' over the 'mass' (considered in Chapter Five), to see the signs of this suppressed intuition that the questions that have the most import and longevity in the humanities are numerical ones. No conception of the sublime has ever been formed otherwise than in geometrical or quantitative terms of ratio and proportion. Whether it is a matter of meaning, feeling or form, questions of number, quantity, magnitude and measure always assert their force, albeit in dim and groping ways. Quantality is to the quantitative as algebra is to arithmetic, or as topology is to geometry, for it is concerned with ratios and not absolute quantities.

Perhaps there is a sense in which the role of the humanities may continue to be phenomenological, in the fundamental sense in which that strain of philosophy has concerned itself not with how things are, but what it is like that they should be that way. A quantical humanities might, as a modest sort of minimum, be concerned with the tonalities of the quantitative. There is no writer who has penetrated further into this enterprise than Sigmund Freud, for whom human drives and feelings were not just illuminated by the kinds of quantality he called the economic, but were entirely unintelligible without it.

Freud is very rarely mentioned in the work of Michel Serres, but he does make a telling appearance in the course of the essay 'The Origin of Language'. The essay announces a unification of biology and psychology through the information theory announced in Claude Shannon's 'A Mathematical Theory of Communication'.[44] Central to Serres' account is the suggestion of a system formed of many levels of integration, in which the coupling of noise and information at one level is integrated into information at the next level up, a process which is describable in the same terms whether one is talking about a cluster of neurons, a cell or a sentence. It is a process in which addition allows subtraction: adding noise creates contradiction, but seeing that contradiction as ambiguity allows for integration, or the subtraction of noise, at a higher level. Sloterdijk's immunology is perhaps a translation of the very same process of turning noise into information. Serres suggests that this Russian doll model is a significant advance on the Freudian 'mechanical or hydrodynamic model',[45] since it allows us to see the unconscious not as a single, turbid reservoir, but rather as an interlocking chain of systems, in which noise at one level is apprehended as information at the next level up:

> At this point the unconscious gives way from below; there are as many unconsciouses in the system as there are integration levels. It is merely a question, in general, of that for which we initially possess no information. It is not a unique black box, but a series of interlocking boxes; and this series is the organism, the body. Each level of information functions as an unconscious for the global level bordering it, as a closed or relatively isolated system in relationship to

> which the noise–information couple, when it crosses the edge, is reversed and which the subsequent system decodes or deciphers.[46]

Information theory provides a way of treating what Serres calls 'the energy account' (the mathematics of thermodynamics, or physical systems in general) and the 'information account' (the mathematics of signals or communication systems communication) as commensurable; they are not on the same scale, but they are of the same order, and therefore capable of being brought into relation.

> The difference between a machine and a living organism is that, for the former, the information account is negligible in relationship to the energy account, whereas, for the latter, both accounts are on the same scale. Henceforth, the theoretical reconciliation between information theory and thermodynamics favors and advocates the practical reconciliation between those funds of knowledge which exploited signs and those which exploited energy displacements.[47]

And, seemingly having left Freud's wheezing, Heath Robinson hydrodynamics behind, Serres suddenly acknowledges that this reconciliation of the energy and information accounts 'was Freud's first dream'.[48] In truth, this particular dream was in fact first and last for Freud, since everything became for him an economic problem. If the early Freud wanted to build a psychophysics of the self, which would translate psychology into measurements of quantity, the Freud of *Beyond the Pleasure Principle* onwards saw this kind of economic endeavour, working in the form of a virtual or performative mathematics, as the very engine of psychological life. Mathematics was not just something in terms of which psychological life could be rendered; it was what it was made of, what it made of itself, in the first place. Freud's theory of the death drive, in its entwining with the erotic or pleasure principle, is not just accidentally but essentially a calculus of thanato-erotic variation and substitution, which is not just expressible as quantitative terms, but brought into being as impassioned quantality. The symbolic relation in which one thing is brought or forced more or less comfortably into correspondence with another is essentially a matter of equivalence or excess, which is to say of measure. This brings psychoanalysis into accord with aesthetics as an interpretative operation.

Freud's work intimates that, if there is the possibility of putting psychodynamics together in the same scale as thermodynamics, then that will formalize a relationship that has been elaborated for centuries in the workings of fantasy calculations and the calculations of fantasy. Revenge and reparation are nothing without the imagination of quantity, and the quantality of imagination, and probably little else beside them. None of our dreams of justice and fairness have any purchase without the aligning and allotting of weights and measures. In fact, the entire teeming universe of qualities is shaped and expressed through scale and extent, ratio and equivalence. Wanting, losing, hating, hurting, enjoying, striving, suffering, knowing, mourning, seeing and saying, all are shaped and suffused by the psychotropic pressure of number. The whole spectrum of affects associated with temporality, its protentions and retentions, its reachings, longings and lastings, is toned and braced through the work of imaginary measure. Chapter Six will show that pleasure itself is not only subject to what Jeremy Bentham called a felicific calculus, but is in itself the application of that calculus. It would be tempting to say that the humanities inhabit the domain of the approximate or imaginary as opposed to what are called the exact sciences, were it not that the idea of exactitude itself is so gravid with quantical fantasy (as is the idea of fantasy). The quantical imagination is a subrealm of the material imagination, and is subject to the reversibility of that phrase; that is, just as the material imagination is both the way in which the idea of matter is imagined, and also the imaginary materialization of the faculty of imagination itself, so the quantical imagination is both the way we dream quantity and the quantical account we keep of our dreamwork. This is double-entry bookkeeping indeed.

We keep ourselves stubbornly and stupidly in the dark about our deep investment, as humanists and humans, in quantality, stumbling about in blind man's buff amid the blaze of noon. If number is part of the process of making explicit, then showing the workings of number in our own rhetorics and forms of cognitive and affective calculus ought to be a more absorbing and impelling task for us all. If we continue to self-medicate with the narcotic conviction that the most essential things we do as humans have nothing to do with number, we not only fail to recognize our best and probably our only prospect for continuing to do valuable things, but reveal that we actually quite literally have no idea what we are doing, for we are all already doing number as much as being it. Undoubtedly, we live in number and also between the implicit and explicit conditions of number. If they are to be good for anything, the humanities must shape up to what I have called quantality – the quality of quantity or the feel for figures – the agitated, affective, philosophical and political imaginary of number.

But taking account of the feeling for number must not bypass the most intense feeling of all provoked by number, namely that of horror.

3

HORROR OF NUMBER

FLAT

Mathematicians like to refer to certain problems of wide implication as 'deep'. It is a quaint word that is rarely heard in other areas, where it has a hint of unearned portentousness. But, of all areas of intellectual activity, mathematics is surely least of all to be characterized in this way. Numbers, the principal constituents of mathematics, are shallow – or, rather, they are absolutely without differentiation as regards their depth.

The curious thing about this flatness is that it is easily and universally apparent, yet almost everywhere resisted. Few people are actually able to bring to bear on numbers the equanimity they appear to enjoin. Most people have favourite numbers towards which they lean. Even if one is indifferent towards numbers, with no preference even for odd over even, the necessary prominence in our lives of certain numbers, the bus I take to work (91), my house-number at school (Middleton B 29), my employee number (355), my PIN (1365), my date of birth, my invariant height and variable weight, seem to light up certain parts of the number line with significance.

Yet there is something unreadable about numbers, if by reading we mean an action of the mind that takes us from something manifest to something else that it indicates or implies, something in whose place it stands. The unreadability of numbers may be intimated by the word 'decipher', for a cipher was originally the name for zero, that number that is not quite one, that yields the name of the unreadability of numbers. Indecipherability seems to signify the particular kind of illegibility that attaches to numbers, that may endlessly be manipulated and moved around, but can never be penetrated.

Though they may denote or measure values, numbers do not have them, or rather they all have precisely the same value, the value of marking some quantity. Qualities are not interchangeable, though they are linked together by relations. Numbers are their relation and nothing more. It is for this reason that Richard Rorty has recommended thinking about numbers as a way to prise ourselves away from the seductions and consolations of essentialist habits of thought. The 'panrelationalism' that he advocates 'is summed up in the suggestion that we think of everything as if it were a number'.[1] Numbers are very hard to think of as having 'an essential core surrounded by a penumbra of accidental relationships'.[2] The essential principle of the number seventeen, say, which happens to be my favourite number, is that it can only be defined relationally. What is more, there are a literally infinite number of ways in which the number seventeen can be defined – as the square root of 289, as the sum of five and twelve, the result of subtracting 5,876 from 5,893 – none of which has any priority over any of the others. A number is nothing more than the sum total, the unsummable total, of all its relations to all the other numbers there might be. For any number, there is no number to which it will not have some relation, and no relation to any number that is more important or intrinsic than any other. Rorty concludes that

> whatever sorts of things may have intrinsic natures, numbers do not . . . it simply does not pay to be an essentialist about numbers. We antiessentialists would like to convince you that it also does not pay to be essentialist about tables, stars, electrons, human beings, academic disciplines, social institutions, or anything else. We suggest that you think of all such objects as resembling numbers in the following respect; there is nothing to be known about them except an initially large, and forever expandable, web of relations to other objects. Everything that can serve as the term of a relation can be dissolved into another set of relations, and so on forever. There are, so to speak, relations all the way down, all the way up, and all the way out in every direction; you never reach something which is not just one more nexus of relations.[3]

Numbers do not have values, because they are the measure of them. Titian's *Diana and Actaeon* is worth £50 million – but what is '50 million' worth? The value of something is that to which it is equivalent, something else into which it may be translated or for which, under certain circumstances, it might be exchanged. This might indeed look like a kind of mathematical operation. Verily, the value of 17 is expressible as the 'something else' of 12 + 5 or 11 + 6. But there really is no else or other in this case, as there is in the case of the price attached to a Titian painting, and for what seems for a moment to be a surprising reason. A Titian can be worth £50 million precisely because a Titian is *not* £50 million, precisely because it can never be fully exchangeable with it. To be sure, I can purchase the Titian for £50 million, but that exchange is not the same as full equivalence, for the previous owner of the Titian cannot treat the £50 million cheque in the same way as the Titian (and if he could the situation would be absurd, since it would amount to exchanging a Titian for another Titian that was identical in every conceivable way, that is, exchanging it for itself). So there can be exchange only where there is non-identity. But this is not the case with the 12 + 5 or 11 + 6 and so infinitely on that constitute 17, precisely because they constitute it. Seventeen is not *worth* 12 + 5, because it is nothing but or other than the fact of its *being* 12 + 5. The alternative expressions for 17 are precisely its identity, even if it is an identity that can never fully be specified, not something that stands in place of that identity. Here, in other words, one really does exchange something for itself, and precisely because it is nothing other than the fact of this exchangeability, and the sum total of all the possible exchanges that will add up to, or subtract down to it.

What is more, there is absolutely no reason to prefer any of these ways of making up 17 to any other. Any one will do just as well as any other for the job of defining 17. This might seem odd in the case of 17, in particular, the seventh prime you get to when you count from 2 upwards. To discover that 17 is a prime is to discover another kind of equivalence for it than the 1 + 16 and the 2 + 15 kind, and the prestige of primes makes it seem as though this is a more important and essential thing about 17 than the numbers of which it is made. In fact 17 is one of only 5 known Fermat primes. A Fermat number is a number such that $F_n = 2^{(2n)} + 1$; 17 is the Fermat number derived from 2. But though these are properties that seem to render it unique, this is not to say any more than that we can pick it out for attention in certain ways, and this is equivalently true for all numbers. Because all numbers are definable in an infinite number of ways, the things that seem to make certain numbers special are just selections from that infinite number of ways. It may seem impressive and mysterious that certain numbers seem to have rare or unique properties, but in fact every number has at least one absolutely unique property – in that it comes between two other numbers in the counting continuum. Every number is generically unique, which means that no number is unique in *being* unique. Uniqueness is what makes numbers so monotonously uniform, not what rescues them from that uniformity. There is no such thing as an uninteresting number, and if there were this would make it interesting in itself.

Picking out certain numbers for special attention is the traditional way of redeeming numbers for human life, because it skews and bunches a system of absolute equivalence into one of differential values, creating a lumpy, striated landscape out of one that is otherwise sleekly, bleakly uniform. Magic numbers, or lucky numbers, suggest that, far from being homogeneous and indifferent, certain numbers do in fact have special qualities or powers – the rule of three, the seventh son of a seventh son, unlucky thirteen and so on. The study of numbers in literature has often depended on this kind of numerological magic. Mathematicians are, perhaps surprisingly, rather drawn to the same kind of mystical or magical properties of numbers – it is as though astronomers were to be drawn to the claims of astrology. One might even say that, in a certain sense, mathematics is a kind of superstitious resistance to the indifference of numbers. A story told by the mathematician G. H. Hardy may bear this out. Hardy had become the patron of the brilliant, self-taught Tamil mathematician Srinivasa Ramanujan, whom he helped to bring to England, where he was elected a Fellow of Trinity. This is the account given by C. P. Snow in his foreword to Hardy's *A Mathematician's Apology* of a visit paid to Ramanujan when the latter was dying in hospital in Putney in 1920:

> Hardy, always inept about introducing a conversation, said, 'I thought the number of my taxi-cab was 1729. It seemed to me rather a dull number.' To which Ramanujan replied: 'No Hardy! No Hardy! It is a very interesting number. It is the smallest number expressible as the sum of two cubes in two different ways' [$1^3 + 12^3$ or $9^3 + 10^3$].[4]

But it may be that it is precisely as the principle of differentiated indifference (numbers are all different from each other in exactly the same ways) that numbers might have the force they do, and this is why my focus in this book is on numbers rather than mathematics. For number is the matter rather than the form of mathematics, and mathematics may be the superstitious deterrence of number. Numbers are what mathematics works on, what it is necessary for there to be in order for mathematical operations to take place. This is perhaps the reason why a purely mathematical definition of number has been so difficult to come up with. Because mathematics thinks with numbers, because it is, precisely, numerical thinking, it has been hard for mathematics, on its own, to think about number. In this sense, although mathematics is made of number, and works in and through it, it is fundamentally opposed to it, precisely because it is the redemption of number. Mathematics, and especially that branch of it known as number theory, seems to show that numbers are not just numbers – that they are tied together by hidden webs of relationship and entailment. By being mathematical, we learn to overlook the most important and defining features of number, namely its flat indifference. It is this feature of number which Lewis Carroll's Red Queen discloses:

> 'Can you do Addition?' the White Queen asked. 'What's one and one and one and one and one and one and one and one and one and one?'
>
> 'I don't know,' said Alice. 'I lost count.'
>
> 'She ca'n't do Addition,' the Red Queen interrupted.[5]

'Doing addition' means being able to process a stream of identical ones into consecutive products, using the embodied adding machine known as counting. This simple action, which seems elementary, is in fact a way of providing orientation, or converting the shallowness of number into a kind of depth. Counting is a way of not 'losing count' amid the swirl of raw numbers.

To be a living entity is to have some kind of here and now, to occupy some particular portion of time and space that can never be merely equivalent to some other portion of time and space. What we call life is perhaps no more or less than this quality of *thisness*, or *itselfness*. It is this thisness that number disperses, flattening it out into equivalence. Number gives control, because number requires and supplies distinctness, the possibility of series and finitude (distinguishability and countability). But it does so at the cost of the drastically asymmetrical, nonreversible world in which my meaning and value is never simply commutable into yours or hers. This absolute equivalence is what I will call death: death, not as nonbeing, but as absolute equivalence, the absence of any difference that would make any real difference between one mode of being and another.

Number is always laying siege to its own numerousness. Numerals fight against the numerous, numbering being an attempt to steady and segment the dizzy delirium of one-after-another, or the one-beside-another, the one-and-another-one-just-like-it. Death is what happens to the one, that which cannot be experienced by anyone else, any other one. As such, death is finitude, the necessity of a limit to being, a limit that allows that being to amount to a singular entity precisely by being cut short. Ultimately, number is death, death being what comes when your number is up. One's death can only ever happen once, and cannot ever be rehearsed or repeated, since it brings to an end the time in which anticipation or retrospection might occur. And yet, it is not one's death that is unique, one's ownmost, in Heidegger's terms, but one's being-towards-death, or dying into death. Death is in fact always the swallowing of the singular by the multiple, the process by which a unique person concurs with the general 'one' of 'one dies.' It is the necessary finitude that gives us our oneness, but that also makes us non-finitely equivalent to every other one who has been born and died.

The horror of number is that any and every number can be counted out as a succession of ones, added to each other (including, of course, decimals and fractions). The units that make up a number are absolutely interchangeable; no one differs in any respect from any other one. And yet those ones are not the same, because they can be added to each other. Counting is a matter of one, then another one, exactly the same, and another one. It is a horror, because it is a vision of indifference – of an absolute differentiation, which makes or is founded on no difference at all, with the number line being a distension of identicality. Indeed, there is evidence that we will not allow ourselves to believe what we claim to know, namely, that the one that is added to eleven to make twelve is the same as the one added to two to make three. Because it is somehow further away, it seems to many that it must be smaller, because distant objects appear smaller than proximate ones.[6] Appalled by the prospect of a flat world that need take no account of the difference we make to it, we tug it into perspective.

HICKORY DICKORY DOCK

There is a tension between numerality and numerology; that is, there is always a tension between the fact that numbers do not and cannot mean anything in themselves, since all numbers must be absolutely equivalent, and the effort to give significance to number. This is no mere philosophical alternation between abstract alternatives. The tension involved is something like that described by Frank Kermode when he points to the inaudibility of 'tick tick', and the need for human beings to give scansion to their experience of time, through transforming the identical series of ticks into the alternating pattern of 'tick-tock'. The sing-song of the tick-tock is a model, says Kermode, of the ways in which we transform the simple one-thing-after-another seriality of abstract time into the meaningful sequences of narrative.

> *Tick* is a humble genesis, *tock* a feeble apocalypse; and *tick-tock* is in any case not much of a plot. We need much larger ones and much more complicated ones if we persist in finding 'what will suffice . . .' Within this organization that which was conceived of as simply successive becomes

> charged with past and future: what was chronos becomes kairos. This is the time of the novelist, a transformation of mere successiveness which has been likened, by writers as different as Forster and Musil, to the experience of love, the erotic consciousness which makes divinely satisfactory sense out of the commonplace person.[7]

So the tension between chronos and kairos is in fact the tension between tension itself and – we seem tellingly not to have a word for this, or not much of one – the untense or the tenseless. Jean-François Lyotard provides a version of this kind of thinking in opposing what he calls the principle of the 'tensor' to a merely semiotic view of signs. The semiotic view considers only relations of equivalence or substitutability between signs, and thus leads to a kind of nihilistic despair, since one will never escape from the play of substitutions. Lyotard accuses semiotics of maintaining a kind of negative economy upon the fact of this lack, which he calls 'the zero of book-keeping'.[8] Against this, the tensor of the sign is thought of as the pure movement of energy, or energy of movement. The tensor is the energy that moves us through and across signs, rather than the attempt to sum or substitute them. In classical Greek, kairos means, not so much a different quality of time, but rather the moment of choice or crisis, which is the moment at which time seems to be gathered into significance. Typically events such as Christian revelation are taken to be the irruption of kairos into the neutral and featureless tick-tick of chronos or merely serial experience (though it is hard to see how there could even be anything like an experience of time as pure seriality).

Very often, almost always perhaps, numerology is involved in this move from chronos to kairos. Temporal significance is established through counting: the countdown to lift-off, the third day, the seventh wave, the millennium. Counting is a reliable and versatile method for achieving orgasm, through the countdown, or in the more mundane version which a tidy-minded girlfriend of my acquaintance once imparted to me, through the mounting consummation of black refuse bags being taken out of the house to the bin. Counting allows us to mark out differences of beat or emphasis; it gives time to prosody by bending it to the force of desire. It locates us in time, making it possible for there to be what we call lived time, by which it seems we must mean regularly interrupted time, time that keeps pausing and resuming, stepping aside from itself, rather than just keeping on keeping on. Numbering is the gateway to this defeat or redemption of pure numerality.

Kairos is associated with revelation, with the suspension of the mundane in favour of the spiritual, or what some contemporary philosophers like thrillingly to call 'the event'. Most theories of revolution borrow from the logic of kairos, in which time is to be redeemed from mere succession. Kairos is thought of as mysterious and fugitive, while the experience of chronos is everywhere. It is, in Kermode's words, 'the interval [that] must be purged of simple chronicity'.[9] (My spellchecker seems to be in league with this mode of thought, for it keeps spontaneously capitalizing kairos as Kairos, while it leaves chronos alone.) But what should be most striking about Kermode's tick-tock principle is that it shows us that it is the experience of kairos that is ordinary and everyday and the experience of chronos that is consequently in need of revelation, precisely because it is so very difficult for us to have any kind of extended experience of the tick-tick that does not start organizing it into sequences. It is not the fact that the organizational structures of *The Divine Comedy* or *War and Peace* or *A la recherche du temps perdu* rescue us from the drab vacuity of the tick-tock that should strike us, but the fact that there is 'humble genesis' and 'feeble apocalypse' even in the tick-tock, that is, right down at the atomic level of our experience of time. The ordinal aspect of numerality, the fact that numbers form an order, is recruited against the cardinal principle of equivalence, the fact that, in order for there to be any meaningful order, each successive number must have exactly the same quality as its predecessor. Ordinality, precisely because it allows for summing and patterning, along with all the apparatus of mathematical thought, is always close to numerology, the magical clumping of differential significance.

These considerations are dramatized in the Marquis de Sade's *120 Days of Sodom*, which offers us a grim, ludicrous accountancy of excess, in which the accountancy is both necessary to the excess and ultimately its antagonist. Not only is the work as a whole constructed in the traditional mode of the story-cycle – the *1001 Nights*, the

Decameron; indeed, the *120 Days* seems to be asking to be read as a Boccaccian *Dodecameron* – its narrative is driven throughout by mathematical procedures. The prominence of number means that the imperious desire for more perversity, which number aims to enforce and augment, always teeters on the brink of tedium and hilarity.[10]

The horror at work in the force of numerality may appear as the inhuman, that which makes human life impossible, or at least unbearable. But in fact, there is nothing more human than this inhuman principle of absolute equivalence, for it seems to be what we bring to nature, which knows no absolute numbers. What is called non-Euclidean geometry, the discovery that Euclidean principles do not apply universally, but only in local worlds like ours constituted according to certain spatio-physical conditions, offers nature an escape from the chill indifference of human mathematics.

Efforts to rescue us from numerality, whether through the ruses of numerology, the rhetorics of redemption or the routines of revolutionary theory, represent themselves as the substitution of quality for quantity, and of force for form. But they fail to recognize the force that is represented in the principle of numerality, and are numb to the fact that the quantitative has a quality and force all of its own. What is more, this quantality, this quantical quality, the quality of being able to be represented as quantity through number, is implicated, indeed, it is an imperative principle, in every attempt to redeem numerality by numerology, since the innumerable will never be able to purge itself entirely of the enumerable. The force of numerality may become visible only rarely 'as such' – indeed, perhaps it is the one-and-one-and-one-and-one of the chronic which ought to qualify as the exotic 'event', rather than the eruption of kairetic significance – but it is the virus within kairetic thinking which gives it all its virulence. Only chronos, the possibility of thinking of moments as moments, which is to say as enumerable units, both distinguishable and equivalent, makes kairos possible.

Much twentieth-century writing, including, in particular, the works of Kafka and Beckett, has attempted to represent or approximate to this kind of anti-kairetic revelation, the revelation of the indefinite condition of finitude, disallowing any kind of lift-off into the infinite (for what is the notion of the infinite but the triumph of number, the eternality of the externality that is counting?), against which all our habits of thought and experience are ranged. We will see it at work in Chapter Nine in the painting of Roman Opałka. Beckett's characters are addicted to counting, but also liable to lose count; indeed the latter is probably the reason for, and consequence of, the former. As for example at the beginning of 'The Expelled':

> There were not many steps. I had counted them a thousand times, both going up and coming down, but the figure has gone from my mind. I have never known whether you should say one with your foot on the sidewalk, two with the following foot on the first step, and so on, or whether the sidewalk shouldn't count. At the top of the steps I fell foul of the same dilemma. In the other direction, I mean from top to bottom, it was the same, the word is not too strong. I did not know where to begin nor where to end, that's the truth of the matter. I arrived therefore at three totally different figures, without ever knowing which of them was right. And when I say that the figure has gone from my mind, I mean that none of the three figures is with me any more, in my mind. It is true that if I were to find, in my mind, where it is certainly to be found, one of these figures, I would find it and it alone, without being able to deduce from it the other two. And even were I to recover two, I would not know the third. No, I would have to find all three, in my mind, in order to know all three. Memories are killing. So you must not think of certain things, of those that are dear to you, or rather you must think of them, for if you don't there is the danger of finding them, in your mind, little by little. That is to say, you must think of them a good while, every day several times a day, until they sink forever in the mud. That's an order.[11]

Mathematicians are less at home in this world than those who think of themselves as unmathematical. Their effort is to generate quality from this grey, toiling mortar of indifference, to build from it a variegated landscape, of pattern, recurrence, contour. And most of all to avoid counting. Solutions that rely on the simple counting out or counting up of possibilities are known by mathematicians as 'brute force solutions', because they do not involve any calculation. There is mathematics in the head of the sunflower because it has discovered, or rather is itself the stochastic precipitate of the discovery, that the most efficient way to pack seeds in a given space is to coil them in a spiral at intervals of .618 of a complete rotation, the Golden Section, or the ratio between successive numbers in the Fibonacci sequence. But the mathematics of the sunflower is immanent to it, not something it can do. Because it is in itself the performance, it cannot perform it for itself. So counting seems unmathematical, because it seems closer to the mathematics that we are than the mathematics we do – hence, perhaps, brutish, *natura naturans* rather than *natura naturata*. Perhaps one answer to the Red Queen's question, what is one and one and one and one . . .? might be *me*, the one that never comes out as one.

Counting is at the heart of mathematical procedure, because every mathematical procedure amounts to, or can come down to, counting. (Whenever one uses an expression like 'every *x*', one is saying that, if one were to count out all the procedures in question, there would be none left over.) And yet mathematics and counting are inimical to each other. We learn to count, which is to say, we train ourselves into a kind of automatism. Counting is never something we can exactly do, precisely because we have in it to give ourselves over to a doing that does itself. The ambivalence of this is noted in Elizabeth Sewell's *The Field of Nonsense*. She argues that

> The Nonsense writer wants to make a world out of language and the mind's pattern of reality, but reality which will be remade so as to be more subject to number; and the characteristics of number and order will have to be imparted to the images in the mind so that they too may be controlled, distinguishable from one another, going along one at a time in an ordered series, limited and exact.[12]

The use of numbers allows for this control because perceptions 'will be brought under stricter control than is usual in language, and in this state they could be played with'.[13] But number involves two associated principles: that of distinctness of units, and that of seriality. Sewell associates them, but there are occasional hints that they may pull in opposite directions, as when she writes that 'as the mention of any number must do, it sets the mind running along the familiar and ordered series of natural numbers. The mind *by the very mention of number is delivered into the hands of its own ordering tendency*.'[14] In Nonsense, the mind is given the freedom to play by a formalization that makes it more the master of the world; but this kind of play also means that one is at a certain risk of being played with by the ordering impulse that is essentially correlative to the impulse to play. Life expresses itself most fully in play, in which death is inevitably recruited.

We can extend this far beyond the writing of Nonsense. The horror of counting is that there is no end to it. Compulsive counters seem to want to make the world controllable by reducing it to number, by making it enumerable. But compulsive counters are also typically themselves compelled by the force of compulsion they attempt to exercise upon the world, as may be seen from James Boswell's account of Samuel Johnson's anxious rituals for negotiating thresholds:

> He had another particularity, of which none of his friends ever ventured to ask an explanation. It appeared to me some superstitious habit, which he had contracted early, and from which he had never called upon his reason to disentangle him. This was his anxious care to go out or in at a door or passage, by a certain number of steps from a certain point, or at least so as that either his right or left foot, (I am not certain which,) should constantly make the first actual movement when he came close to the door or passage. Thus I conjecture: for I have, upon innumerable occasions, observed him suddenly stop, and then seem to count his steps with a deep earnestness; and when he had neglected or gone wrong in this sort of magical movement, I have seen him go back again, put himself in a proper posture to begin the ceremony, and, having gone through it, break from his abstraction, walk briskly on, and join his companion.[15]

Boswell's description makes it clear that counting is never enough, since counting must itself be subject to checking, which is to say, recounting. And, when counting itself becomes subject to the force of counting, where is any end to be made, how is one ever to cross the threshold into a result?

LESS THAN ONE

Counting means adding one to one to one to one. One adds one, then adds another one, then another. But what is one of something? Who has ever truly seen one of anything? There can be one only once there is two; but then there can never be one again. One is the quality of oneness that any one thing has in common with another thing that can be counted as one. This is Bertrand Russell's definition of a number, as 'the set of those classes that are similar to a given class', where 'similar to' means having the same number of elements.[16] One must be able to count two things as one for either one of them to count as one. In counting, one is never a complete unit, for it depends upon the possibility of there being another one. Unless there could be another one to be added to any one, it would remain less than one, come up short of the one that it must nevertheless be taken to be in order for counting to ensue. You must just assume you know what one is in order to add another one to it, even though only that addition will confirm the oneness of the one. One completes itself in the sundering into two that will keep it eternally at a distance from itself.

There are two modalities of the one. There is the oneness of indivisibility, that in which no difference of parts may be discerned. Such a one is not countable as one, precisely because there is nothing that does not belong to it. This, presumably, is the condition of the Lord God, brooding in his self-belonging prior to the creation, or the oneness of the universe, which must by definition include everything. Then there is the one that counts as one, the one that may be counted off, that is one as the first unit in a sequence, meaning that it is one that is seen from the viewpoint of two or more, the more that two means. This is perhaps the reason why the Greeks did not count one as a number. To make the One of the Allness of the cosmos into a one, it must be mutilated into unity. If there really were only the One, if there were no real division in the universe, then that universe would not yet be thinkable of as a one, for it would not be possible to step outside it to count to two. One must always be more than one in order to avoid being less than one. One is always a more-than-one that is less than one, for counting will never let you get the one to add up exactly to itself.

Problems that can only be solved by brute force, such as the varieties of the Travelling Salesman Problem, or TSP, which involves finding the shortest itinerary between any sequence of points that will involve visiting each point once and only once, are regarded as mathematically insoluble. The fact that ant colonies are able to solve this kind of problem by trial and error, and that artificial ant colonies may be generated to do it, does not make the solutions any more mathematical, because they are all blind, and it is this blindness that accounts for the horror of merely counting. Mathematics is the sidestepping or recoil from this horror.

Horror, because horror is at its heartless heart uncountability. Freud suggests that the head of the Medusa signifies castration, not just because it is decapitated, but in 'confirmation of the technical rule according to which a multiplication of penis symbols signifies castration'.[17] No mention of this 'technical rule' appears to my knowledge anywhere else in Freud, which suggests a nonce-rule, or a once-rule, a rule which is more-than-one (as all rules have to be), yet also less-than-one (since it is a rule applied and probably invented only for this occasion, like Rule Forty-two in Alice's court).[18] Norman O. Brown remarks that, in this world where the many can stand both for the less than one and the more than one that protects against it, 'we are in a world oscillating between the one and the many, a world of fission and fusion, the world of schizophrenia; the world of the schizophrenic patient whose "primary function in life, as he saw it, was to restore people who had been *multilated*".'[19] Horror refers to the sensation of bristling, in which the skin, that organic avatar of the integer, may lose its integrity, standing up as though multiplied into hairs, or shivering, shuddering or quaking. Horror is dispersal amid horripilated multiplicity, a dissolution into the innumerable. Horror is simply losing count.

Counting up and counting out are defences against this horror of indifference, but they also threaten to expose us to it. Counting opens us up to the one-by-one of every composition, the material substrate of every relation. It delivers us to the very dread of losing count from which it preserves us. Noel Carroll's *Philosophy of Horror* suggests that horror is a reaction to forms of categorical incompleteness or indistinctness.[20] It might seem as though the horror of number contradicts this, since number is the very principle of distinctness. But there is a special horror in the indistinctness that results from the pseudo-distinctiveness of number. A quantity is definite and absolute, it may seem to have an aura or quality about it. But it is made up of the joining of a certain number of indistinct units, indistinct because they have to be accumulations of the absolutely identical, of an entity added inexorably to itself. There can be little doubt that the horror of insects has to do with their multiplicity, with the sense of unbearable, spawning multitude they provoke – if they crawl over us, they induce the sensation of our skin itself, as we say, crawling. The categorial uncertainty of the spider surely has something to do with the fact that we are not sure how many legs it has (eight is just beyond the limit of our power to grasp a number without needing to count it). Knowing that it has eight and not six is a sort of protection against the sheer uncountability of its legs. Added to this is the horror that the fly or spider seems indifferent to the subtraction of one or more of its legs. So spiders and flies embody the horror of number because they seem so indifferent to the very numbers that differentiate them. Numbering is a superstitious protection against number. No millipede has a thousand legs, though some have as many as 750. I have heard that it is impossible for a centipede to have a hundred legs, but for a reason that, for some reason, I find truly appalling, namely that centipedes *always have an odd number* of pairs of legs.

Most cultures distinguish odd from even numbers, and many see odd numbers as luckier or more powerful than even.[21] The Pythagoreans believed that odd numbers were male and 'lordly'.[22] This belief is still at work in Shakespeare's Falstaff, who, in *The Merry Wives of Windsor*, declares that 'there is divinity in odd numbers, either in nativity, chance or death', and in Cleopatra, who laments, on hearing of the death of Antony, that 'the odds is gone/ And there is nothing left remarkable/ Beneath the visiting moon', where 'the odds' seems to mean that which stands out savingly from the lunar churn of mere recurrence.[23] The odd seems to have become identified more recently with the principle of the absurd, the residual or the unfinished, precisely because it seems to nudge us on into a bad infinity of uncompletable succession, a temporality in which we can never succeed in being on time. Why else is it that when you awaken disorientated in the night, the time shown on a digital clock always seems to be an odd number – 1.29, or 3.13?

Counting belongs to both sides of number, to the formed and the formless, the discontinuous and the endlessly ongoing. But counting always exposes one to the chance of losing count. We teach children to count to protect them against the horror, claustrophobic and agoraphobic at once, of the one and one and one and one of the countless or uncountable.

And, if one way of responding to the horror of number's indifference is to seek forms of concentration, another is to seek to identify with radiation. This is to set the distributive against the distinctive. Being is necessarily locative; it belongs to deixis, here, now, this. But there are those who identify not with the where and when of being, but with the fields of distributed relations within which being clusters and condenses: not with position, but with disposition. Not 'I am painting,' or 'this is my painting,' but 'there is painting.' Some of these have found in the field called art a way of not identifying with identity, but with the fields within which identity is figurable, with the ground against which the figure stands out. It need not be art, and insisting on art's priority in this is a failure of nerve. It is just that art offers some ways of doing what may be done otherwise. The doodler knows a minor form of this radiative drive, the desire to fill space with relations, the desire not to be in space, but to be it. The artist is identified with the brush, the stylus, the mark with which one makes one's mark. But the field of art has been attractive to some, in the way that mathematics is to others, because it can sometimes also allow the possibility of diffusing oneself across the system of relations, not concentrating oneself at the tip of the brush, but inhabiting the entire apparatus, hand, eye, canvas. The saturation of space and the emptying of space are here equivalent, because both are equivalence itself. The preposition that governs this kind of work, this work that works against the standing out of work, is *across*. The plain must become the plane.

The ideology of number in the modern world is that number is inhumanly exact, while the realm of the word, the tone, the gesture, is vitally imprecise. We are many of us still spontaneous Bergsonians in this respect, favouring the fuzzy continuities of the temporal against the harsh, anonymous, mechanical, severing pseudo-exactness of the spatial. The phrase 'the exact sciences' sums up the difference between the realms of the inhuman mathematical and the human. But the mathematical is not the realm of number. The exposure to pure number, or pure exposure to number, exacts a kind of horror or delirium, which does not belong to either side of the exact/inexact equation, precisely because it is not equal to anything, not even itself.

Literary writing will often incorporate the idea of number as exactness, in order to immunize itself against it, in order to keep the equation balanced between the mathematical–mechanical and the unmathematical organic, in order to assert the powers of life over the deathliness of number. But the deathliness of number is also a strange life-in-death. Freud's account of the way in which the death instinct routes itself through life is a dramatization of this. Death wishes to be the zero that answers and balances the one that is life. But in order to do this, life must be made to add up to one – death that supervened upon a life that was not yet a life would not be death. So death must put itself off, must start a count through life that will never come to an end, since being alive means losing count. To die exactly is impossible, because there are too many things to be 'lived off', as Freud weirdly, brilliantly puts it, of which account would have to be taken.[24] The failure to be one, the certainty of losing count in the more-than-one that will always be less-than-one, is what puts number, the very domain of the deathly, on the side of life, as the indefinite, as the not-yet-finished, the 'unnullable least' or 'leastmost all', as Beckett puts it, that can never be reached.[25] The fact that numbers can never properly add up to anything, that number can never fully come to an end, is what makes number so deathly and yet allies it with a kind of craving agitation that can come close to rapture. This is the delirious horror of number.

4

MODERN MEASURES

'WHAT HAVE I TO DO WITH MILLIONS?'

Writers and literary critics have often thought of themselves as representative of a rival estate to number, that of the word, and have been keen to assert the ways in which literary texts may have offered forms of exception or demur from social and political pressures to extend quantification and calculative rationality. Literature is represented as providing a growling, and occasionally, as in a novel like Dickens's *Hard Times*, a clamorous *non placet* to such norms. Increasingly during the nineteenth century, literature sees it as its role to respond to and resist the empire of number. Accounting and calculation become the opposite of the kinds of human experience to which the arts, and especially the literary arts, give expression. As the dominion of number grows, so does the assumption that it is the vocation of literature (and, for some, of art in general, but literature has a favoured role) to embody the values of the singular and the particular against the mass, the mean, the normalizing aggregate, and to resist the growing ascendancy of the quantitative over the qualitative. This anti-numerical ideology is articulated by Jane Eyre, when she complains bitterly of her public denunciation as a liar, and refuses Helen's consolation:

> 'Everybody, Jane? Why, there are only eighty people who have heard you called so, and the world contains hundreds of millions.'
>
> 'But what have I to do with millions? The eighty, I know, despise me.'[1]

George Eliot, who thought more seriously than most other nineteenth-century novelists about questions of scale and quantity, also represents this point of view, in Lydgate's reflections on entering the medical profession: 'Considering that statistics had not yet embraced a calculation as to the number of ignorant or canting doctors which absolutely must exist in the teeth of all changes, it seemed to Lydgate that a change in the units was the most direct mode of changing the numbers.'[2] In fact, from the nineteenth century onwards, the ways in which art and literature would seek to secede from the republic of number began to take necessarily numerical forms – as literature sometimes represented itself as the guardian of the one against the many, the moiety against the mass, and sometimes as the asserter of the infinite against the finite.

Ralph Waldo Emerson was among those who were most attentive to the onward march of number. Like many others, from Hegel onwards, he found an image for the coordination of the singular and the multiple in the figure of Napoleon, writing that 'if any man is found to carry with him the power and affections of vast numbers, if Napoleon is France, if Napoleon is Europe, it is because the people whom he sways are little Napoleons.'[3] Napoleon is representative of the 'living labor' of democratic capitalists, as opposed to the 'dead labor' of conservatives whose capital is locked up massively but inertly in land and buildings:

> The first class is timid, selfish, illiberal, hating innovation, and continually losing numbers by death. The second class is selfish also, encroaching, bold, self-relying, always outnumbering the other and recruiting its numbers every hour by births. It desires to keep open every avenue to the competition of all, and to multiply avenues: the class of business men in America, in England, in France and throughout Europe; the class of industry and skill. Napoleon is its representative.[4]

Emerson represents these two classes mathematically, in terms of an unbalanced sum, with stagnation in one column and increase in the other, but, in a sense, he is really counterposing two modes of number: number as mere quantity and number raised, so to speak, to the power of numerousness. These masses of little Napoleons are not just manifold, but, in their powers of expansion, represent the accretive power of number itself. Emerson opposes intensity of experience to the calculation of quantity. Writing of religious experience in 'The Over-soul', he says

> Our religion vulgarly stands on numbers of believers. Whenever the appeal is made, – no matter how indirectly, – to numbers, proclamation is then and there made that religion is not. He that finds God a sweet enveloping thought to him never counts his company . . . It makes no difference whether the appeal is to numbers or to one. The faith that stands on authority is not faith. The reliance on authority measures the decline of religion, the withdrawal of the soul.[5]

The same principle of non-quantifiable intensity is articulated in Emerson's 'The Transcendentalist': 'It is the quality of the moment, not the number of days, of events, or of actors, that imports.'[6] The injunction to seize the day is an old and frequently renewed one, of course; but the framing of such an injunction, not as a snatching of some precious thing from oblivion, but as a defiance of the homogeneity seemingly induced by number-consciousness, seems a distinctively nineteenth-century gesture. And yet it is also characteristic of this defiance that it must borrow from what it breaks from. In 'Experience', Emerson writes that

> To finish the moment, to find the journey's end in every step of the road, to live the greatest number of good hours, is wisdom. It is not the part of men, but of fanatics, or of mathematicians, if you will, to say, that, the shortness of life considered, it is not worth caring whether for so short a duration we were sprawling in want, or sitting high. Since our office is with moments, let us husband them. Five minutes of today are worth as much to me, as five minutes in the next millennium.[7]

The statistician and the political economist are blamed for reducing everything to equivalence and thereby emptying everything of value. But standing out against the principle of equivalence enjoins another kind of mathematics; for Emerson does not counsel any simple abandonment to the passing moment, but rather a vigilant accountancy of 'the greatest number of good hours'. Emerson's anti-quantitative calculus is repeated in Walter Pater's urging, some 25 years later, that, with 'a counted number of pulses' only being our mortal allowance, we learn to 'pass most swiftly from point to point, and be present always at the locus where the greatest number of vital forces unite in their purest energy'.[8] Since we inhabit only a brief interval of time, and 'our one chance lies in expanding that interval, in getting as many pulsations as possible into the given time', only an intensive sensory calculus will permit us to resist our finitude.[9]

It is highly characteristic of efforts to substitute quality for quantity from the nineteenth century onwards that the path beyond number should seem always to have to pass through it. Matthew Arnold complained in his *Culture and Anarchy* (1869) of the tendency for socially reforming ideas to get swallowed up in bureaucracy:

> an English law . . . is ruled by no clear idea about the citizen's claim and the State's duty, but has, in compensation, a mass of minute mechanical details about the number of members on a school-committee, and how many shall be a quorum, and how they shall be summoned, and how often they shall meet.[10]

But in discerning what Arnold called 'the intelligible law of things', and in offering the claim that literature should protect the particular, the anomalous and the minute, such a criticism did not so much reject number as implicitly prefer small numbers to large, hence the quality of elective minority that has characterized modern literary and cultural self-definitions. Arnold declared in *Culture and Anarchy*:

> when we speak of ourselves as divided into Barbarians, Philistines, and Populace, we must be understood always to imply that within each of these classes there are a certain

> number of aliens, if we may so call them, – persons who are mainly led, not by their class spirit, but by a general humane spirit, by the love of human perfection; and that this number is capable of being diminished or augmented. I mean, the number of those who will succeed in developing this happy instinct will be greater or smaller, in proportion both to the force of the original instinct within them, and to the hindrance or encouragement which it meets from without.[11]

Following Arnold, F. R. Leavis suggested in his 1930 pamphlet *Mass Civilization and Minority Culture* that

> A reader who grew up with Wordsworth moved among a limited set of signals (so to speak): the variety was not overwhelming. So he was able to acquire discrimination as he went along. But the modern is exposed to a concourse of signals so bewildering in their variety and number that, unless he is specially gifted or especially favoured, he can hardly begin to discriminate. Here we have the plight of culture in general. The landmarks have shifted, multiplied and crowded upon one another, the distinctions and dividing lines have blurred away, the boundaries are gone.[12]

It should not be too surprising, in a book the title of which so openly endorses the principle of ratio in its critical reasoning, to find an economic metaphor at its head: what Leavis calls the 'accepted valuations' of a culture 'are a kind of paper currency based upon a very small proportion of gold'.[13]

During the nineteenth century, the conditions of production and circulation of literary texts became ever more, and more conspicuously, bound up with large numbers. As literature became a mass phenomenon, the second half of the nineteenth century saw, not just an avalanche of numbers, but a secondary landslide of words, in huge numbers. Not only did literature become a mass-market phenomenon, it also, like so much else in our fantasy of the Victorian, itself became massive (there is a certain rite of passage undergone by every student of English who is tempted to trumpet the alleged fact that Victorian novels are so long and loquacious because Victorian authors were paid by the word). In other words, literature became, as it had always been of course, but now more copiously and conspicuously, and as part of its own self-understanding, quantitative – much of a muchness.

So, while literature found itself opposed to the order of number, it also found itself entering into it, meaning that there was a deep participation of number in writing and writing in number from the middle of the nineteenth century onwards. But this is not just a question for literature, or is a question for literature precisely because it is such a general question. Literary writing is ever more suffused with numbers and numerical awareness, because number, in the prodigious variety of its aspects and occasions, entered indissociably into so many forms of modern experience.

The later nineteenth century saw the beginnings of a curiosity about the nature of the perception of number, or the inner life of numerical consciousness. In the early 1880s, Francis Galton, that great, obsessive counter of everything, from the number of shocks he received on a train journey from London to Liverpool to the number of fidgets per minute in a bored audience, became interested in the subjective processes attached to number itself.[14] His work in the early 1880s was driven by two interlocking preoccupations: the 'psychometric' idea of measuring and quantifying mental processes, and the mental processes attaching to quantity and measure. Along with essays on the statistics of imagination, Galton also published work on the visualization of numbers. He even toyed with the possibility of conducting arithmetic by smell.[15]

Number did not simply assault experience, it penetrated and transformed it. At the inauguration of what would become known as phenomenology, or the philosophy that dealt with the way things appeared to the mind, separate from the quasi-mathematical questions of what was true, real or coherent, there is Edmund Husserl's first book, *Philosophy of Arithmetic*, an expansion of his Habilitationsschrift of 1887, 'On the Concept of Number' ('Uber den Begriff der Zahl'), a work criticized by Frege for its emphasis on psychology rather than logic.[16]

From the late nineteenth century onwards, literary writing began to move ever closer to number. In certain areas, notably in France and in Russia, a quasi-mystical mathematical poetics asserted itself, especially in writers like Mallarmé and Valéry in France. In Britain and Germany, mathematics itself slowly began to change its character over the nineteenth century. In the first part of the century, number could still be thought of as what Mary Poovey calls the 'modern fact': 'modern facts are assumed to reflect things that actually exist, and they are recorded in a language that seems transparent. Since the early nineteenth century, this transparent language has been epitomized by numerical representation.'[17] Over the course of the nineteenth century, mathematics became ever more formal, more relational, more projective and speculative in its forms. That is to say, it became ever more abstracted from the visible and verifiable facts of physical and social life. Non-Euclidean and *n*-dimensional geometries became a staple of late nineteenth-century supernaturalism. If, in the first half of the nineteenth century, numbers appeared to lock down indefiniteness, in the second half, the emphasis was increasingly on variation, frequency and fields of distribution. Instead of the quantification of uncertainty, the second half of the nineteenth century saw something like the virtualizing of number. A certain number mania emerges intelligibly from this inversion in early twentieth-century writing.

For a modernist writer, it might have seemed self-evident that the realm of number was becoming ever more abstract, autonomous and powerful; ever more firmly and ever more grimly set over against what modern writers began to think of as the 'life-world'. In a century of mass slaughter, number comes in fact to be the name of death itself. As is often the case, this apprehension fails to account for the most striking features of numerical and calculative awareness, and fails so totally that one must suspect that it does so in order precisely to defend against them, and to defend in particular against the acknowledgement of the ways in which the life of number has both proliferated and transformed from the nineteenth century onwards. Rather than locking the life-world up in an iron cage of calculative rationality, number begins to thread through social and personal life as intricately as the vascular system.

SCALE

As social relations were expressed more and more in number, representations of number tended to stress its alienness. Numbers came to stand for the inhuman, the mechanical, the unconscious, the impersonal, the inert. W. H. Auden's poem 'Numbers and Faces' articulates what might be called the humanist ideology of number. I do not have much fondness for the word *ideology*, which is often just a way of describing somebody else's ideas in such a way as to represent them not as ideas at all but as a kind of mental illness – but in this case, the systematic preference for the unsystematic that is expressed in an opposition to number does seem to make the term appropriate. Auden finds those of a numerological persuasion, who suggest that life is governed by the mystic force of certain 'small numbers', merely 'benignly potty', in contrast with the totalitarian statisticians of mass existence, who 'go horridly mad'.[18]

Against the obsessive–compulsive wielders of number, either at the minor, neurotic scale, or on a major, despotic scale, Auden promises the ultimate unquantifiability of human relations signalled by the non-numerical uniqueness of the face, 'for calling/ Infinity a number does not make it one.'[19]

Modern writers, and the critics who formed the climate in which they lived, moved and had their being, tended to conflate the realm of number with the fact of large numbers, which they identified with blurring, conformity and standardization of response. The liberal view of quantification is expressed by E. M. Forster in his novel *Howards End*, in the person of Ernst Schlegel:

> 'It is the vice of a vulgar mind to be thrilled by bigness, to think that a thousand square miles are a thousand times more wonderful than one square mile, and that a million square miles are almost the same as heaven. That is not imagination. No, it kills it. When their poets over here try to celebrate bigness they are dead at once, and naturally.'[20]

There was often a sinister underside to this in the evocation of the unnumberable mass to be found in the nightmares of liberal intellectuals and anti-democrats alike, as laid out so chillingly in John Carey's *The Intellectuals and the Masses*.[21] We might find this enacted in D. H. Lawrence's beastly poem 'How Beastly the Bourgeois Is':

> Let him meet a new emotion, let him be faced with another
> man's need,
> let him come home to a bit of moral difficulty, let life face
> him with a new demand on his understanding
> and then watch him go soggy, like a wet meringue.
> Watch him turn into a mess, either a fool or a bully.[22]

Lawrence's own reaction, when faced with this 'new demand on his understanding', is a collapse of his own into homicidal sogginess:

> How beastly the bourgeois is!
>
> Standing in their thousands, these appearances,
> in damp England
> what a pity they can't all be kicked over
> like sickening toadstools, and left to melt back, swiftly
>
> into the soil of England.[23]

In 'Let the Dead Bury Their Dead', the dead 'are in myriads' – not so much because there are myriads of them, but because multiplicity is death itself:

> The dead in their seething minds
> have phosphorescent teeming white words
> of putrescent wisdom and sapience that subtly stinks;
> don't ever believe them.
>
> The dead are in myriads, they seem mighty.
> They make trains chuff, motor-cars titter, ships lurch,
> mills grind on and on,
> and keep you in millions at the mills, sightless pale slaves,
> pretending these are the mills of God.[24]

Numbers like 'thousands' and 'millions' here become the very bearers of unimaginability. The nescient hordes of the unimaginative are mathematically generalized into the unimaginable, enabling them to be swept away all as one and all at once.

Not that Lawrence was always opposed to the dominion of number, provided that the numbers were small and precise enough, as becomes clear in 'Tortoise Shell', his meditation on the cruciform patternings of a tortoise's body in the volume *Birds, Beasts and Flowers* of 1923:

> It needed Pythagoras to see life playing with counters
>   on the living back
> Of the baby tortoise;
> Life establishing the first eternal mathematical tablet . . .
>
> Turn him on his back,
> The kicking little beetle,
> And there again, on his shell-tender, earth-touching belly,
> The long cleavage of division, upright of the eternal cross
> And on either side count five,
> On each side, two above, on each side, two below
> The dark bar horizontal.
>   The Cross![25]

Lawrence was the modernist writer who mounted the most sustained assault against the realm of number, determined as he was to assert quality over quantity, the hazy, nebular, indefinite or indistinct, which is said to be living, over the exact and numerable world, which is said to be abstract, mechanical and dead (or male for short). In the words of Birkin in *Women in Love*, this modulates into an aristocratic ideology of the absolute nonrelativity of value:

> 'We're all the same in point of number. But spiritually there is pure difference and neither equality nor inequality counts. It is upon these two bits of knowledge that you must found a state. Your democracy is an absolute lie – your brotherhood of man is a pure falsity, if you apply it further than the mathematical abstraction.'[26]

In fact, however, Lawrence's writing, so relentlessly pitted against number, is in fact more strangely mesmerized by it than almost any other modern writer (I have a feeling that there is a higher number of number words to be found in his work, especially big-number words, like hundreds, thousands and millions, than in most other modern writers). There is, for example, Gudrun, thinking about Gerald's mechanism:

> The wheel-barrow – the one humble wheel – the unit of the firm. Then the cart, with two wheels; then the truck, with four; then the donkey-engine, with eight, then the winding-engine, with sixteen, and so on, till it came to the miner, with a thousand wheels, and then the electrician, with three thousand, and the underground manager, with twenty thousand, and the general manager with a hundred thousand little wheels working away to complete his make-up, and then Gerald, with a million wheels and cogs and axles.
>
> Poor Gerald, such a lot of little wheels to his make-up! He was more intricate than a chronometer-watch. But oh heavens, what weariness! What weariness, God above! A chronometer-watch – a beetle – her soul fainted with utter ennui, from the thought. So many wheels to count and consider and calculate! Enough, enough – there was an end to man's capacity for complications, even. Or perhaps there was no end.[27]

Two antagonistic principles are locked together here, as they perhaps are in every articulation of the ideology of number. The first is the idea of Gerald's mechanism, which consists in reducing life to distinct and countable units, to produce order, control and predictability. But this very exactness produces the intolerable indistinctness of the indifferently non-identical, the 'madness of dead mechanical monotony and meaninglessness'. In its account of Gudrun's meditations, Lawrence's narration becomes a kind of counting-up procedure, going up in squares, 1, 2, 4, 16, but precisely in order to be able to open on to the uncountable 'so on', the losing count that lurks in every counting procedure.

What is there to set against this? There are the various forms of intensity, the modernist moments of being, the radiant epiphanies, the absolutely singular and incommensurable 'events'. In *Women in Love*, there is Mrs Crich, who 'lost more and more count of the world, she seemed rapt in some glittering abstraction, almost purely unconscious' – though 'she bore many children.' And there is Hermione, in her consuming fantasy of murdering Birkin:

> A terrible voluptuous thrill ran down her arms – she was going to know her voluptuous consummation. Her arms quivered and were strong, immeasurably and irresistibly strong. What delight, what delight in strength, what delirium of pleasure! She was going to have her consummation of voluptuous ecstasy at last . . . She lifted her arm high to aim once more, straight down on the head that lay dazed on the table. She must smash it, it must be smashed before her ecstasy was consummated, fulfilled for ever. A thousand lives, a thousand deaths mattered nothing now, only the fulfilment of this perfect ecstasy.[28]

Moments of intensity must be rescued from the nightmare of monotonous numbering, but the logic that does the rescuing must nevertheless be numerical, first, in that it derives from and depends upon the rounding up or counting-as-one of the idea of the 'mathematical' or the 'mechanical', and, second, in that it must assert itself as a transcendent one, now conceived as an absolute, and entirely impossible, equality to itself. It is a transcendent counting-as-one of that which transcends counting altogether, but is really an apotheosis of the number one, as though there could be a one that bore no relation to any other kind of singularity. But a singularity that bore no relation to any other singularity would not be anything. Lawrence retreats from numerical horror, in which the ones can never be got to add up into a nameable, numerable total, into numerical fantasy, the idea of a one that must always hover asymptotically below the threshold of the dead consummation of being a one comparable to any other one (the horror of democracy). There is an erosion of the one at either end of the scale, the elementary and the ultimate.

Numeration is the deterrence of this erosion, this failure of the one to stand up and be counted. Ecstasy, epiphany, the event, are the singular without seriality, a consummation that can never be summed up, a uniqueness that goes beyond or refuses to be cashed in as the merely unitary, which is the nullity of the one-like-another-one.

Lawrence is enraged at the capacity of number to resist numeration, its capacity to make us lose count. He declares in 'Bestwood', 'What we should strive for is life and the beauty of aliveness, imagination, awareness, and contact. To be perfectly alive is to be immortal.'[29] Lawrence sees quantity as death itself – but acting on the impulse to preserve life against death actually requires the most brutal entry of all into the quantitative, in the form of social culling: 'I know that we should look after the quality of life, not the quantity. Hopeless life should be put to sleep, the idiots and the hopeless sick and the true criminal. And the birth-rate should be controlled.'[30]

A sinking feeling assails liberal-authoritarian modernists, in the sudden, horrified sense that the world might indeed be full of individual minds, of which the measure might never be able to be taken. The preference for the small over the great, as we find it articulated for example in Virginia Woolf, goes along with a certain desire for accumulation, or enlargement of scale and number – 'One wanted fifty pairs of eyes to see with, she reflected. Fifty pairs of eyes were not enough to get round that one woman with, she thought.'[31] Modernism flees, not from number as such, but from large numbers into small numbers. Though modern literature and culture may try to get themselves on the other side of number, the very obsession with this anumerical project makes modern writing quantical at every turn.

This seems to mirror the actual leaning, both of mathematical reasoning, and of the technical and engineering work based upon it towards a sensitivity to very small numbers. Prompted by his discussion of the importance of aluminium in the period he characterized as that of 'neotechnics' as opposed to 'paleotechnics' (heavy industry) of the previous century, Lewis Mumford wrote in his *Technics and Civilization* (1934) that

> the significance of minute quantities – which we shall note again in physiology and medicine – is characteristic of the entire metallurgy and technics of the new phase. One might say, for dramatic emphasis, that paleotechnics regarded only the figures to the left of the decimal, whereas neotechnics is preoccupied with those to the right.[32]

We may see this even earlier. The year 1900 saw the appearance of two works that, following the number-magic of date-coincidence, may be regarded as reciprocally illuminating. Sigmund Freud's *The Interpretation of Dreams* announced the method of psychoanalysis, a method which depended upon the isolation and amplification of tiny and seemingly insignificant phenomena of mental life, and Max Planck's formulation of the radiation law determined that the radiation emitted by a hypothetical black body (a theoretically perfect absorber of radiant energy) must be emitted in discrete packets or quanta, each of them multiples of the value known thereafter as Planck's Constant. Quantum physics is so called because it is built on Planck's discovery that the world is not completely continuous at the smallest scales. At these scales physical actions cannot take an infinite number of values. Rather, those actions must be multiples of a particular quantity. It is as though the world turned out to be cubist at its core – the more one turned up the resolution, the more blocky or granular it appeared to be. And it is precisely this granularity that accounts for many of the disturbing features of quantum mechanics, in which particles are not permitted to move smoothly and continuously from one state to another, as at higher scales, which smooth out those spikes and jumps into continuous lines, but rather must jump between conditions.

For both Freud and Planck, large significance inheres in tiny variations detectable only by close and minute analysis. Both Freud and Planck announce a world in which, as Virginia Woolf was to say influentially in her essay 'Modern Fiction' (1921), we should not 'take it for granted that life exists more fully in what is commonly thought big than in what is commonly thought small'.[33] In fact, modernist literature and criticism participates in what may be called the scale-commutation that is characteristic of modern science, whereby small local fluctuations are amplified to have very large effects. Virginia Woolf is typical in the large, rather booming claims she tends to make in defence of the minute and the particular: as Kim Shirkhani has written, 'Woolf discredits analytical, abstract statements even as she herself dispatches them.'[34] The importance of the atom, and of even smaller particles, is not so much their smallness as their mathematical tractability, the fact that they moved, following the work of Maxwell, Boltzmann, Planck and others, from the realm of hypothesis into the realm of number and calculation. The sentence in 'Modern Fiction' before the one I have just quoted in which Woolf asks for an amplified attention to the small enjoins: 'Let us record the atoms as they fall upon the mind in the order in which they fall, let us trace the pattern, however disconnected and incoherent in appearance, which each sight or incident scores upon the consciousness.'[35]

Woolf's writing in fact evokes the communication between the very large and the very small, and asks some of the same questions about the mathematics of the very small and the very large as mathematicians asked. Often this involves reflection on the idea of vibrations, with which there had been a general intoxication in art and literature from the late nineteenth century onwards. *Mrs Dalloway* evokes the slight yet huge perturbation of a magisterial car driving up Bond Street:

> The car had gone, but it had left a slight ripple which flowed through glove shops and hat shops and tailors' shops on both sides of Bond Street. For thirty seconds all heads were inclined the same way – to the window. Choosing a pair of gloves – should they be to the elbow or above it, lemon or pale grey? – ladies stopped; when the sentence was finished something had happened. Something so trifling in single instances that no mathematical instrument, though capable of transmitting shocks in China, could register the vibration; yet in its fullness rather formidable and in its common appeal emotional; for in all the hat shops and tailors' shops strangers looked at each other and thought of the dead; of the flag; of Empire.[36]

We are told of a hypothetical instrument of infinitesimal sensitivity, not in order to discredit the notion of unconscious sensation, but in order to validate it, by giving it a plausible correlate in the physics of very small quantities. A similar kind of scale-commutation occurs in the description of a First World War air raid in Woolf's *The Years* (1937):

> A gun boomed again. This time there was a bark in its boom.
> 'Hampstead,' said Nicholas. He took out his watch. The silence was profound. Nothing happened. Eleanor looked at the blocks of stone arched over their heads. She noticed a spider's web in one corner. Another gun boomed. A sigh of air rushed up with it. It was right on top of them this time . . . The Germans must be overhead now. She felt a curious heaviness on top of her head. One, two, three, four, she counted, looking up at the greenish-grey stone. Then there was a violent crack of sound, like the split of lightning in the sky. The spider's web oscillated.[37]

Modernist writing is characterized, not by the eschewal of systems of calculation and enumeration – of time, money, people – but the interest in the ways in which such systems could be converted into each other. It is an interest, not in the units, but in the exchanges between systems of units. This accounts for the interest in counting to be found throughout the work of Joyce, Beckett, Lawrence, Woolf, Sinclair and many others. For all these writers, counting is an indispensable way into the marking out of syncopations, or complex, crossed rhythms.

Like many other modernists who devoted themselves to the making out of other kinds of rhythms than those measured by the clock, Woolf is closely attuned to the work of Henri Bergson, whose doctoral dissertation, published in English as *Time and Free Will* (1910), offers a critique of the idea that sensations have anything at all to do with number. Sensations are registered in terms of variable intensity in time, argues Bergson, while number relates to extension, that is, to magnitudes juxtaposed in space. Bergson's book is an extended critique of the 'psychophysics' of the late nineteenth century, as epitomized by the quantitative views of sensation introduced by Ernst Weber and Gustav Fechner, in particular the Weber–Fechner law, which proposes that the intensity of a sensation is proportional to the logarithm of the stimulus intensity. Bergson concluded that

> in consciousness we find states which succeed, without being distinguished from one another; and in space simultaneities which, without succeeding, are distinguished from one another, in the sense that one has ceased to exist when the other appears. Outside us, mutual externality without succession; within us, succession without mutual externality.[38]

Bergson sees number as a reduction of experience to merely spatial relations. But it is Bergson himself who is guilty of the reduction, in his oddly archaic, even arthritic, imagination of space, and his reduction of the operations of number to operations in and on space. Here, it seems, mathematics can only play its part in a critique of reductiveness if it has itself been plausibly but brutally reduced.

MEAN

Modernist writing often strives to display its excess over number, but also feels impelled to guard against the spilling exorbitance that is characteristic of number. Accordingly, modernist writers often embrace the values of leanness and niggardly spareness – in versions of what Joyce called the art of 'scrupulous meanness' of his *Dubliners*.[39] The term 'meanness' bulges oddly with alternative meanings. It suggests that being mean with language may give more meaning, or may focus attention purely on the meant. It may imply niggardness or parsimony, or a focus on the smallness of the lives it discusses. But it may also imply a kind of writing that hovers around the mean, the intermediate or the average. One obsolete Old English meaning of *mean* is fellowship or sexual intercourse, deriving from aphetic Old English *gemǣne*, that which is shared or common, similar to the Old Frisian *mēne* meaning assembly. There is a shimmer of imprecision in the modernist mean.

May Sinclair's *Life and Death of Harriett Frean* appeared in 1922, the *annus mirabilis* of modernist experiment. The novel is concerned with a life lived in self-sacrifice. It delivers a punitive measure for measure in the matching of its mode to the self-denying structure of Harriett's life. The nursery-rhyme opening suggests comparisons with James Joyce's *A Portrait of the Artist as a Young Man*:

> 'Pussycat, Pussycat, where have you been?'
> 'I've been to London, to see the Queen.'
> 'Pussycat, pussycat, what did you there?'
> 'I caught a little mouse under the chair.'
>
> Her mother said it three times. And each time the Baby Harriett laughed. The sound of her laugh was so funny that she laughed again at that; she kept on laughing, with shriller and shriller squeals.[40]

But, where Joyce's novel shows a growth into style, Harriett Frean's narration allows scarcely any progression beyond this pinched equivalence. Of course, in one sense, the novel intends us to read it obliquely – to shrink from and protest against its mode, as we recoil from the self-disavowing ethic. It gives us short measure in order to provoke the angry and assertive hunger for more that Harriett Frean learns to subdue. But it also depends upon the very exactitude of accounting that it mocks and mourns. Its subject, of course, is denied the power to read her own life in these terms. By the end of her life, exhausted and dimmed by the effort of self-repression, Harriett is unable even to do the household accounts: 'Her head dropped, drowsy, giddy over the week's accounts. She gave up even the semblance of her housekeeping' (HF, 170).

There is a calculus of disease counted out remorselessly in the novel that precisely lines up with the disease of calculation:

> Months passed, years passed, going each one a little quicker than the last. And Harriett was thirty-nine.
>
> One evening, coming out of church, her mother fainted. That was the beginning of her illness, February, eighteen

> eighty-three. First came the long months of weakness; then the months and months of sickness; then the pain; the pain she had been hiding, that she couldn't hide any more. (HF, 99–100)

So the novel sets risk, adventure, love, self-assertion against the keeping of strict moral accounts. But risk and adventure are just as disastrous in the novel as prudent self-restraint. At its heart is the catastrophic financial speculation of Harriett's adored stockbroker father:

> 'There's nothing gross and material about stock-broking. It's like pure mathematics. You're dealing in abstractions, ideal values, all the time. You calculate – in curves.' His hand, holding the unlit cigar, drew a curve, a long graceful one, in mid-air. 'You know what's going to happen all the time.
>
> '. . . The excitement begins when you don't quite know and you risk it; when it's getting dangerous.
>
> '. . . The higher mathematics of the game. If you can afford them; if you haven't a wife and family – I can see the fascination . . .' (HF, 38)

Harriett's father not only goes in for reckless financial speculation, the 'higher mathematics of the game', he also reads dangerous, speculative books, by Darwin, Huxley and Spencer, all the time impelled by '"fascination in seeing how far you can go"' (HF, 41). But this is an intellectual improvidence that reduces, not only Frean's own family, but the friends who follow his advice. The novel insists on the same remorseless code of equivalence that it condemns. Self-repression will lead inevitably to hysteria, whether it takes the form of the hysterical paralysis of Harriett's friend Priscilla, or the outbreak of delirium that Harriett seems to suffer under the anaesthetic during her cancer surgery. Moral restraint involves the keeping of books, the mirroring repetition of the same: 'They sat side by side at the dinner table and in school, black head and golden brown leaning to each other over the same book; they walked side by side in the packed procession, going two by two' (HF, 30). But Harriett cannot follow these books. She attempts to follow her father in his reckless intellectual speculation:

> She made a point of finishing every book she had begun, for her pride couldn't bear being beaten. Her head grew hot and heavy: she read the same sentences over and over again; they had no meaning; she couldn't understand a single word of Herbert Spencer. He had beaten her. As she put the book back in its place she said to herself: 'I mustn't. If I go on, if I get to the interesting part I may lose my faith.' (HF, 43)

In the end, Sinclair's *Life and Death of Harriett Frean* appears to consume itself in poverty, swallowing up, with its exact balancing of accounts, even the ironic supplement of awareness that would allow one to see and condemn its narrowness. And yet the watchful equilibrium it maintains requires a paradoxically spilling excess of meanness, in which, with words reduced to obsessive numbering, numbers threaten to overtake words, in the counting out, counting down of time, age, money, hunger and desire that ticks and clicks throughout the novel:

> She knew what she would have. She would begin with a bun, and go on through two sorts of jam to Madeira cake, and end with raspberries and cream (HF, 12) . . . She passed through her fourteenth year sedately, to the sound of Evangeline (HF, 26) . . . they walked side by side in the packed procession, going two by two (HF, 30) . . . Year after year the same. Her mother parted her hair into two sleek wings (HF, 49) . . . Two, three, five years passed, with a perceptible acceleration, and Harriett was now thirty (HF, 67) . . . Eighteen seventy-nine: it was the year her father lost his money. Harriett was nearly thirty-five (HF, 82) . . . by the spring of eighteen-eighty he was upstairs in his room, too ill to be moved (HF, 87) . . . Harriett wondered why he was making that queer grating and coughing noise. Three times (HF, 92) . . . Mr Hichens had given them six weeks. They had to decide where they would go (HF, 93) . . . Months passed, years passed, going

> each one a little quicker than the last. And Harriett was thirty-nine (HF, 99) . . . There was one chance for her in a hundred if they had Sir James Pargeter: one chance (HF, 100–101) . . . she had died in agony, so that she, Harriett, might keep her hundred pounds (HF, 107) . . . She was forty-five, her face was lined and pitted and her hair was dust colour, streaked with gray (HF, 119) . . . 'Whatever happens to Beatie he must have his sweetbread, and his soup at eleven and his tea at five in the morning' (HF, 130) . . . Fifty-five. Sixty. In her sixty-second year Harriett had her first bad illness (HF, 161) . . . The years passed: the sixty-third, sixty-fourth; sixty-fifth; their monotony mitigated by long spells of torpor and the sheer rapidity of time (HF, 173) . . . Three more years. Harriett was sixty-eight (HF, 176).

Words here shrivel down to numbers, which threaten reciprocally to swell uncontrollably beyond the constraining powers of verbal language.

CONTINUITY AND DISCONTINUITY

The modern world experiences itself in terms of speed and flux, of a *glissade* that outruns perception. But modernity is also, coincidentally and consequentially, a matter of measurement. This is often represented as a conflict between the quality of 'pure' movement, announced (Bergson) and denounced (Wyndham Lewis) throughout modern literature and culture, and efforts to measure and calibrate that movement. In fact, as with many dichotomies, these two alternatives provoke and perpetuate each other. Modernity is expressed and experienced as fluent speed, to be sure; but it is also embodied and epitomized in the speedometer – a word that is recorded in print for the first time in *The Times* in 1904, offering this instrument as an optional extra on the Ford Model T, first produced in 1908. Modern thinking is toned and textured by the pull between quality and quantity, intensity and measure – in short, between the continuous and the discontinuous. We might associate the speedometer, as emblematic modernist device, with the switch, which gave to the modern world its characteristic capacity for abrupt and absolute transitions between on and off, slow and fast. It is striking that Samuel Beckett, who had devoted so much attention in his work to slow diminishments and gradual fadings out, should have left, as the final words of the last play he ever completed, 'Make sense who may. I switch off.'[41]

This tension between continuity and discontinuity expresses itself in a heightened awareness of the defining role of scale in resolving molecular or corpuscular aggregates into lines or series. We might pause to recall what Michel Serres repeatedly notes, that the derivation of *tension* and allied words like *tone* and *tune* is uncertain: perhaps from *teino*, to stretch, but perhaps also from *temno*, to cut or dissect. The same ambiguity attaches to the word *rhythm*, which comes from the Greek *rhein*, meaning to flow, even as rhythm is precisely that which chops duration into measures.[42] Modernist movement quakes with these tremors of minimality. Beneath or within the blur of persistence of vision, there is the 'dynamite of the tenth of a second', splitting or parcelling out every apparent continuity.[43] We may note, as a prelude to the ideas laid out here, the primal preoccupation of cinema with the fact of explosion. Moving image was not just employed to analyse into their components the gestures and movements of human bodies, it was also employed to restore or rearticulate the temporal contour of an action that seemed to consist of nothing but dissolution.

Leopold Bloom puzzles over one version of this dichotomy in the Sirens episode of *Ulysses*.

> Numbers it is. All music when you come to think. Two multiplied by two divided by half is twice one. Vibrations: chords those are. One plus two plus six is seven. Do anything you like with figures juggling. Always find out this equal to that. Symmetry under a cemetery wall. He doesn't see my mourning. Callous: all for his own gut. Musemathematics. And you think you're listening to the etherial. But suppose you said it like: Martha, seven times nine minus x is thirtyfive thousand. Fall quite flat. It's on account of the sounds it is.[44]

Students continue to call this kind of thing 'stream of consciousness' and to be subject to head-shaking reproof for it; for, with Bloom indeed, there is very little quality of the stream, with his typically staccato shunts of thought from one idea to the next, sometimes juddering forward along a more or less intelligible straight line, sometimes swerving aside, as in the sudden reflections on Richie Goulding's purblind appetite. Perhaps the only example of streaming in his choppy discourse is the coinage 'Musemathematics', and even this lacks the collideorscape inventiveness of many of Joyce's later word-blends. 'Musemathematics' is a slight improvement on 'mathematical music', since it seems to activate the idea of musing, anticipating the 'museyroom' of *Finnegans Wake*, but the improvement is minimal.[45] The word wobbles a little between the singular of 'music' and the plural of 'mathematics', recalling the slight incongruity of the phrase 'Numbers it is,' an incongruity precisely of what is known grammatically as 'number'.

Indeed, we might view all the 'music' of the 'Sirens' episode as the kind of mathematization about which Bloom speculates here, involving as it does the addition, subtraction and transposition of letters and words, considered as quanta. 'Sirens' is full of lyric lengthenings of vowels which may appear to make the words croon and yearn ('Seabloom, greaseabloom'), but it also has reduction to musical elements, like the scale picked out in relaying the actions of the deaf waiter: 'Bald deaf Pat brought quite flat pad ink. Pat set with ink pen quite flat pad. Pat took plate dish knife fork. Pat went.'[46] At times, the writing of the chapter itself seems to adopt Bloom's mathematical reading of music, in which the mimicry of musical form actually bleaches out musical effect from the words, which are reduced to what Garrett Stewart has called 'alphabetic integers'.[47] In this respect, the chapter itself can seem as tone-deaf as Pat the waiter. An example would be Anthony Burgess's reading of the phrase 'Blmstdp' for 'Bloom stood up' as a mimicry of a 'hollow fifth', a chord in which the thirds are suppressed.[48] This identification benefits from the mathematical pun which makes the (notoriously ugly or unmusical) fifth the result of the removal of the vowels, of which there are five, both because there are five vowels and because five of them (o, o, o, o, u) have been removed.

But we should note that Bloom's musings go in opposite directions. 'Numbers it is. All music when you come to think' could mean both 'Music is just a matter of numbers when it comes down to it,' or the more Pythagorean 'all mathematical relations are a kind of music.' The reference to 'the etherial' might also evoke Pythagoras, as well as the coalescence of matter and movement in the magical, all-pervading pseudo-substance the ether, which Lord Salisbury, in an address to the British Association for the Advancement of Science, described as 'the substantive case of the verb "to undulate"'.[49]

Bloom's apprehension of music as mathematics is embodied in the bit of apparatus that, elsewhere in *Ulysses*, Ned Lambert demonstrates for indicating which turn is on in the music hall. Joyce reminds us of this apparatus in a wonderful conjoining of letters and numbers, as we briefly glimpse Blazes Boylan's secretary putting aside her book at work:

> Miss Dunne hid the Capel street library copy of *The Woman in White* far back in her drawer and rolled a sheet of gaudy notepaper into her typewriter.
>
> Too much mystery business in it. Is he in love with that one, Marion? Change it and get another by Mary Cecil Haye.
>
> The disk shot down the groove, wobbled a while, ceased and ogled them: six.
>
> Miss Dunne clicked on the keyboard:
>
> — 16 June 1904.
>
> Five tallwhitehatted sandwichmen between Monypeny's corner and the slab where Wolfe Tone's statue was not, eeled themselves turning H. E. L. Y.'S and plodded back as they had come.[50]

Bloom's mathematical musings recall Helmholtz's demonstrations of the compound nature of sound vibrations in *On the Sensations of Tone*, and come at the end of a long period of reflections on the atomistic components of composite forms.[51] The nineteenth century had seen the invention and popularization of the idea of the mathematical curve, in the characteristic bell shape of Gaussian or normal distribution. More generally, it had made familiar the idea that the movement or transmission of messages or information was most conveniently effected by decomposing and then reassembling the information to be transmitted. The ideal of maximal convertibility and maximum communicability developed in the nineteenth century – which might be seen as an ideal of maximal continuity, or minimized discontinuity – required the intervention of a discontinuous medium. We have become used to the observation that the séance and later on the telepathic soirée were a kind of parallel to technological forms of transmission like the telegraph, for they employed precisely the same logic of decomposition followed by synthesis. Table-rapping was a sort of slow typewriting. In this sense, all the allegedly analogue technologies of the nineteenth century were in fact digital, or protodigital, in that they involved the translation of a continuously variable wave into a discontinuous set of variations. The epitome of the protodigital medium was the cinema, with its dramatic slicing and dicing into the continuity of the visible for the purposes of capturing and reproducing motion. The most widespread digital code in the nineteenth century was Morse, which makes its appearance in *Ulysses* in Stephen's evocation of the dog on the shore of Sandymount: 'His snout lifted barked at the wavenoise, herds of seamorse. They serpented towards his feet, curling, unfurling many crests, every ninth, breaking, plashing, from far, from farther out, waves and waves.'[52] The passage indicates not only the oscillations of position that make up each individual wave, but also, at a higher level, the alternating current of the waves, which allows them to be distinguished and enumerated. These are oscillations that are not only measurable as number, but oscillations in and out of the condition *of* number.

ZENO

A simple way of summarizing this concern is to say that modernity encounters in a series of strikingly practical ways the paradoxes of Zeno regarding movement. In accounts of modernism, Bergson has carried the day against Zeno. This is partly because the Bergsonian critique of the conjoined values of calculation, quantity and intellect and his correlative praise of pure becoming have become sovereign in our ways of thinking about art and its relation to time and meaning.

The time philosophy of Bergson drew together into one current the preference for the power of intuition, which was able to apprehend the complex persistingness in fluidity and fluidity in persistingness of things over the power of intellect, which was said to divide the world up into static objects. Bertrand Russell characterizes this attitude neatly and tartly in his essay 'The Philosophy of Bergson' of 1912: 'Thus logic and mathematics do not represent a positive spiritual effort . . . but mere somnambulism, in which the will is suspended, and the mind is no longer active. Incapacity for mathematics is therefore a sign of grace – fortunately a very common one.'[53] We have taught ourselves to prefer motion conceived, not as the magical and, for Zeno, inconceivable passage from one fixed condition to another, but as the pulsive current of becoming, in which nothing is ever entirely left behind, and all is a continuous commingling of retention and protention.

For Bergson, the force of 'life' was not only immeasurable, it was also inexhaustible. This was good news for the Allies in the First World War, who, Bergson assured his readers in 1914, were fighting an enemy fuelled only by its own barbarous mechanism, whereas the Allies were on the undefeatable side of life: 'to the force which feeds only on its own brutality we are opposing that which seeks outside and above itself a principle of life and renovation. While the one is gradually spending itself, the other is continually remaking itself.'[54] It is not just that Germany would gradually run out of nitrates and international credit; it is that it had blunderingly elected to measure itself by measurability itself.

Measurability, and in particular the infinite divisibility proposed in Zeno's paradoxes, belongs to an order of discontinuity to which Bergson opposes an absolute continuity. Ultimately Bergson's opponent is divisibility itself. To a world of distinct objects, Bergson opposes a world of unresting energies and commingled vibrations. The distinction between two colours is a distinction between two frequencies, which seems absolute only because of 'the narrow duration into which are contracted the billions of vibrations which they execute in one of our moments'.[55] Slow down our perceptions so that we can 'live it out at a slower rhythm',[56] and we are synchronized with those vibrations, and the colours blur together, or become simply vibrations, as when one pushes one's nose right up to a Seurat or a pixellated video screen, a kind of experience which, oddly enough, Bergson refers to, not as pure quantity, but as 'quality itself', the blending of sensation and perception.[57] Like many others, Bergson finds in the new physics, and especially particle physics, a contradiction of the fundamental discontinuity in nature announced by Democritus, in his announcement that all that exists are atoms and the spaces between them.

> We see force more and more materialized, the atom more and more idealized, the two terms converging towards a common limit and the universe thus recovering its continuity . . . the nearer we draw to the ultimate elements of matter the better we note the vanishing of that discontinuity which our senses perceived on the surface.[58]

Bergsonism amounts to the demand for 'the idea of an universal continuity' – the smooth enjambment effected by that 'n' before 'universal' being a bit of phono-philosophical confirmation offered by Bergson's translators.[59] The same actions of magnification and deceleration are involved in Walter Benjamin's decimal and decimating dynamite, the effect of which was to explode the seeming integrity of forms. Dance and music provide Bergson with examples absolute continuity, in which before and after are indisseverable:

> If jerky movements are wanting in grace, the reason is that each of them is self-sufficient and does not announce those which are to follow. If curves are more graceful than broken lines, the reason is that, while a curved line changes its direction at every moment, every new direction is indicated in the preceding one. Thus the perception of ease in motion passes over into the pleasure of mastering the flow of time and of holding the future in the present.[60]

At the end of the third chapter of his *Creative Evolution*, Bergson provides us with an image of the movement of history, as an immense and all-inclusive push of life through matter:

> As the smallest grain of dust is bound up with our entire solar system, drawn along with it in that undivided movement of descent which is materiality itself, so all organized beings, from the humblest to the highest, from the first origins of life to the time in which we are, and in all places as in all times, do but evidence a single impulsion, the inverse of the movement of matter, and in itself indivisible. All the living hold together, and all yield to the same tremendous push. The animal takes its stand on the plant, man bestrides animality, and the whole of humanity, in space and in time, is one immense army galloping beside and before and behind each of us in an overwhelming charge able to beat down every resistance and clear the most formidable obstacles, perhaps even death.[61]

Bertrand Russell comments sourly on this passage that

> a cool critic, who feels himself a mere spectator, perhaps an unsympathetic spectator, of the charge in which man is mounted upon animality, may be inclined to think that calm and careful thought is hardly compatible with this form of exercise. When he is told that thought is a mere means of action, the mere impulse to avoid obstacles in the field, he may feel that such a view is becoming in a cavalry officer, but not in a philosopher, whose business, after all, is with thought.[62]

For Bergson, intellect does violence to the world by fragmenting it; for Russell, Bergsonian intuition does violence to the world by forcing distinctness into unity, cramming multiplicity together into an imperious, expanding One. This would in fact be the result of Bergson's absolute continuity without discontinuity, in which time only ever thickened. Without the possibility of discontinuity, in fact, without the numerable divisions insisted on by Zeno, there would be continuity, but no movement; there would be, we may suppose, movement, but no internal spacing or differentiation, no movement away, no self-distancing of time, which could therefore do no more than convolute and coagulate, like the weirdly smeared state of matter known as the Bose–Einstein condensate, occurring at temperatures close to zero, that was predicted in 1924–5 and experimentally produced in 1995.[63]

There is no absolute discontinuity between continuity and discontinuity. Zeno shows, not that movement is impossible, but rather that it may not be possible to make movement fully intelligible. He is right, not because there is no movement, but because movement is not intelligible without dependence on a principle of discontinuity that seems to make it impossible. The problem posed by the Eleatic paradoxes is not how to explain movement, but how to explain the fact that movement has these two incompatible but indispensable dimensions, one merely aggregative, the other integral. Perhaps the reason why the mind is so exercised by the question of motion is because, in considering it, it must itself alternate, itself thus 'mined with a motion', between the plenitude of the continuous and the ration of the discontinuous.[64] Readings of the paradoxes like Bergson's are an attempt to create an absolute discontinuity between the discontinuity of intellect and the continuity revealed to intuition. Meanwhile the conditions of modern life continued to confirm the existence of Zeno's paradox, confirming that there was in fact no discontinuity between continuity and discontinuity.

ALTERNATING CURRENTS

Continuity and discontinuity are the two polarities between which much modern writing alternates. Indeed, we might say that this is enacted in the principle of alternation itself, as it became more and more apparent in the machineries of the modern world. This is the period in which the concepts of periodicity and frequency enter powerfully, if also subliminally, into general awareness, uniting the discontinuous and the continuous at a higher level, since, in the repeating wave, the continuously variable curve forms a series of distinguishable oscillations. The movements of modernism are as much pulsations as propulsions. They are also movements of active, but immobile, agitation. In a sense, modernity comes into being with the defeat of Thomas Edison's direct current system of electricity and the adoption of the Tesla–Westinghouse system of alternating current. From now on, vectors would be indissolubly joined to oscillations, those inner atoms, or elementary particles of movement. Modern movement was the movement of fans, flywheels, propellers, dynamos, escalators, revolving doors (patented by Theophilus van Kannel in 1888) and rotary engines of all kinds, Gatling guns, cranked cameras and projectors, the tank, which propelled itself forward by means of repeated rotations of a single repeating loop, and helical instruments such as the phonograph and, later, the tape recorder. Their operation brought into being a whole new dinning tinnitus of sounds – hummings, whirrings, rattlings, hissings, whizzings and buzzings. Cinema turned the binary alternations of instruments like the thaumatrope into a progressive form, but this happened by degrees, and Lynda Nead has observed how common it was for early films, especially striptease films, to be shown repeatedly backwards and forwards.[65]

If it is true that modernism has a defining attraction to the continuously varying waveform and indeed to curves of all kinds, it is also often drawn to the exposure of the blocky elements of those curves, through slowings or close-ups that reveal the elements of which they are aggregated. Garrett Stewart has pointed to what he calls the 'flicker effect' of modernism, in which the spool seems to get caught or to judder, suddenly revealing the individual components of which film or narrative is composed, moments at which the numberless becomes suddenly numerable.[66] A literary parallel to this is to be found in Conrad's *The Secret Agent*, in which we read of Winnie Verloc's perplexity at the fact that her mother suddenly starts spending half-crowns and five shillingses on cab fares. Her mother's unaccustomed 'mania for locomotion' is explained when she reveals that she has secured new accommodation in an almshouse 'founded by a wealthy innkeeper for the destitute widows of the trade'.[67] Conrad renders her last cab journey as an extraordinary kind of agitation on the spot:

> In the narrow streets the progress of the journey was made sensible to those within by the near fronts of the houses gliding past slowly and shakily, with a great rattle and jingling

> of glass, as if about to collapse behind the cab; and the infirm horse, with the harness hung over his sharp backbone flapping very loose about his thighs, appeared to be dancing mincingly on his toes with infinite patience. Later on, in the wider space of Whitehall, all visual evidences of motion became imperceptible. The rattle and jingle of glass went on indefinitely in front of the long Treasury building – and time itself seemed to stand still.[68]

TELLING

A writer is often most palpably and painfully quartered between quantity and quality during the process of writing. Even before the days of automatic word counts, writers were driven to keep the count on what they are writing, in words, paragraphs, chapters and pages, and later print runs, sales figures and revenues, for all that they might struggle to forge more fluent and organic forms of unity. Few writers have documented this process as evocatively as Virginia Woolf in her diaries, the diary being a form of writing which is cast between the uncaring ordinality of the clock and the more fluctuant and approximate graphings of thought and feeling. There is a sometimes bathetic counterpoint in Woolf's diaries between the attempts to capture fleeting insights and state of feeling and the rendering of calendrical accounts, through the recording of times, dates, birthdays. She begins her first entry for 1919 explaining that she is restricted by a hand injury to one hour's writing a day, but that 'having hoarded it this morning I may spend part of it now, since L. is out and I am much behindhand with the month of January.'[69] With her 37th birthday a few days away, on 25 January 1919, she imagines what will be her attitude at the age of fifty to what she will find written there: 'If Virginia Woolf at the age of 50, when she sits down to build her memoirs out of these books, is unable to make a phrase as it should be made, I can only condole with her' (WD, 17). She undertakes, for the benefit of the 'elderly lady' (WD, 17) she will be at fifty, to provide a full account of her friends, their achievements and a forecast of their future works, concluding that 'The lady of 50 will be able to say how near to the truth I come; but I have written enough for tonight (only 15 minutes I see)' (WD, 18). A week short of thirteen years later, with *The Waves* behind her and her fiftieth birthday in sight, Woolf parcelled out for herself, alas too generously, a prospectus of the two decades of work still in front of her:

> I shall be fifty on 25th, Monday week that is: and sometimes I feel I have lived 250 years already, and sometimes that I am still the youngest person on the omnibus (Nessa says that she still always thinks this as she sits down.) And I want to write another four novels: *Waves*, I mean; and *The Tap on the Door*; and to go through English literature like a string through cheese, or rather like some industrious insect, eating its way from book to book, from Chaucer to Lawrence. This is a programme, considering my slowness, and how I get slower, thicker, more intolerant of the fling and the rush, to last out my 20 years, if I have them. (WD, 174)

There are interferences in these periodic cycles; later, she notes that it would have been her father's 96th birthday, and that if he were still alive she would never have written anything. Even her mood is given an implicit calibration, with her references to the fluctuations of her 'spiritual temperature' (WD, 265). While she was writing, she oscillated between the desire for fluency, or continuity, and the desire for a kind of compacted integrity. Her sense of the book she is writing is alternately light, floating, fluent, dashing and then 'tense and packed',[70] 'a hard muscular book' (WD, 102). Time and again, she speaks of her desire to smooth out 'chop and change',[71] to create a kind of unity. These two conditions are imaged in the move from handwriting to typescript. She struggles to find the right kind of composite image for her own composition. Often, this image is of some entity made up of oscillation itself, for example *The Moths*, the idea of which 'hovers somewhere at the back of my brain' (WD, 131). The title of the book itself shuttles for a few weeks between *The Moths* and *The Waves*, which we can read as a kind of alternation between different alternating frequencies, the rapid whirr of the moth's wing, and the slow, booming or thudding pulse of the waves – in '*the concussion of the waves breaking . . . with muffled thuds, like logs falling, on the shore*'.[72] Throughout the diaries, the activity of thought is rendered in the quasi-mechanical whirrings and flutterings of insects: 'The mind is the most capricious of insects – flitting, fluttering' (WD, 124). Woolf strives to achieve fluency, but also is brought repeatedly to the recognition that this continuity is only achievable against the background of resistance formed by the steady whittlings of the clock: 'I find myself in the old driving whirlwind of writing against time. Have I ever written with it?' (WD, 127). This sensation of mobility achieved against resistance precipitates a marvellous, self-referring image of rooks suspended, windhover- or angel-like, in the blustery air:

> I have to watch the rooks beating up against the wind, which is high, and still I say to myself instinctively, 'What's the phrase for that?' and try to make more and more vivid the roughness of the air current and the tremor of the rook's wing slicing as if the air were full of ridges and ripples and roughness. They rise and sink, up and down, as if the exercise rubbed and braced them like swimmers in rough water. (WD, 131)

Woolf's diaries are full of actions of counting up, counting out and counting off, and they provide an energizing metre throughout her novels too. There is Susan in *The Waves* grimly counting down the days to the end of the school term: 'I count each step as I mount, counting each step something done with. So each night I tear off the old day from the calendar, and screw it tight into a ball.'[73] Then there is the tense counting-off of the seconds to track the progress of the German air raid in *The Years*, in the passage from which I quoted earlier:

> Nicholas looked at his watch as if he were timing the guns. There was something queer about him, Eleanor thought; medical, priestly? He wore a seal that hung down from his watch-chain. The number on the box opposite was 1397. She noticed everything. The Germans must be overhead now. She felt a curious heaviness on top of her head. One, two,

> three, four, she counted, looking up at the greenish-grey stone. Then there was a violent crack of sound, like the split of lightning in the sky. The spider's web oscillated.
>
> 'On top of us,' said Nicholas, looking up. They all looked up . . .
>
> One, two, three four, Eleanor counted. The spider's web was swaying. That stone may fall, she thought, fixing a certain stone with her eyes. Then a gun boomed again. It was fainter – further away.
>
> 'That's over,' said Nicholas. He shut his watch with a click.[74]

As Joyce's washerwomen in 'Anna Livia Plurabelle' assert, 'every telling has a taling,' and there is no writer in whose work the actions of narration and numeration are more entrained with each other than Samuel Beckett. Molloy's account of his laborious communication with his deaf and blind mother brings together sequence and repetition, cardinality and ordinality, in telling fashion:

> I got into communication with her by rapping on her skull. One knock meant yes, two no, three I don't know, four, money, five goodbye. I was hard put to it to ram this code into her ruined and frantic understanding, but I did it in the end. That she should confuse yes, no, I don't know and goodbye were all the same to me, I confused them myself. But that she should associate the four knocks with anything else but money was to be avoided at all costs. During the period of training therefore, as I administered the four knocks to her skull, I would stuff a banknote under her nose, or in her mouth. In the innocence of my heart. For she seemed to have lost, if not all notion of mensuration, at least the capacity of counting beyond two. It was too far for her. The distance was too great from one to four. By the time she reached the fourth knock she imagined she was still at the second, the first two having been erased from her memory as completely as though they had never been felt.[75]

The counterposed orders of narrative and re-counting, of narrative and counting, here give the passage its characteristically syncopated movement. The fling and lingering of the writing are cross-cut with the table-rapping of the séance and the clopping of Clever Hans's hooves. We can perhaps think of the knocking on Mag's skull as the whirrings of Woolf's inspirational moths or waves slowed down to a frequency at which the cycles become countable. Just when it appears we are about to be allowed to break out of Mag's own autistic binarism, and the unintelligible knockings and bangings are about to build into a kind of narrative *durée*, we are dragged back to the order of elementary bodily percussion – the solution to Molloy's difficulty being not the invention of another code, but the simple amplification of the old one:

> I looked for and finally found a more effective way of putting the idea of money into her head. This consisted in replacing the four knocks of my index knuckle by one or more (according to my needs) thumps of the fist, on her skull. That she understood.[76]

Like the fabled monoglot Englishman whose method of getting a foreigner to understand him is to Speak More Loudly, this is the introversion of sense (in the double French sense of meaning and direction), the turning of movement on and into itself, to form a standing wave, a thrumming, mobile matter of pure pulsation.

Mathematics, and in particular number, are therefore far from being the adversaries of grace and intuition, are in fact levers and accelerators of the modernist evocations of speed, flux and the desire for 'universal continuity'.[77] Number is important because it forms part of the alternation between orders of magnitude and the principles of continuity and discontinuity, melody and percussion, principles which come together in the heightened awareness of states of flicker, fluctuation and alternation, and in a focus on the atomistic divisibility of forms of modern movement, depending as they often do on mechanical operations involving multitudinous but innumerable repeated processes. Modernist movement is mathematized: in it, matter is riddled with motion and motion condensed and accelerated into a tensely tremulous kind of matter. Modernity may be characterized by the multiplicity of the gearing mechanisms needed to effect these transpositions between levels, scales and ratios. The epistemological apparatus required for this might be given the same name as that invented at the beginning of his career by the poet, Marxist literary critic and sometime engineer Christopher Caudwell – the automatic infinitely variable gear.[78]

5

LOTS

'Mony a mickle', tradition and consonance between them assure us, 'maks a muckle.' 'Maks', or 'mak'? The point at which singular becomes plural or plural becomes singular will be the point of this chapter, which will be concerned with the characteristics of what may be called the 'quasi-choate'. Lots of little things make up a lot. Most languages abound in words for these bundled abundances, these muchnesses. It is a necessary characteristic of such aggregate terms that they are defined vaguely, in terms of more or less, rather than precise quantities. But this is by no means to say that they are not numerical, for the intuition of number will have a great deal to do with their definition and use.

Sometimes these aggregate-terms harden into quasi-units of measurement. We frequently hear nowadays of a 'shed load' of something, usually meaning an impressively and unexpectedly large amount, often of some undesired or unwelcome commodity. The term was originally employed, and is still sometimes so employed, to refer to the contents of a lorry that have been accidentally spilled, or 'shed', on to a road. More and more, however, one senses that the people using the expression see the shed as an adjectival noun rather than a past participle, indicating a certain volume of material, that is, as much as could be contained within a typical garden shed. This in its turn vaguely summons up images of sheds being towed up and down the motorway system, occasionally disgorging their cargoes of trowels, garden gnomes and hosepipes. The subsequent mutation of 'shed load' into 'shit load' is perhaps assisted by the fact that excrement tends to occur in piles or heaps, which, though they vary considerably, share the characteristic that there is usually too much of them, or more of them than is needed.

We may think of these terms as vague, but the imprecise quantities or arrangements they name can in fact be quite precisely delineated. Language exhibits surprising nuances when it comes to distinguishing between different kinds of aggregate. A nuance, after all, is a shading or a clouding; though a cloud is itself a more-or-less aggregate, there are so many different kinds of cloud that, until the systems for grouping and distinguishing them were developed at the beginning of the nineteenth century, they might have appeared as the figure of crowding itself. A heap, for instance, is usually thought of as both bigger and less organized than a pile, since piled things are more regular in their construction than heaps. Piles have been piled: heaps may come into being spontaneously. So dirty laundry comes in heaps rather than piles – a pile of shirts would have to be folded neatly, and a pile of anything is likely to consist of examples of the same class of thing. There is a certain zone of overlap where a pile of objects can reasonably also be referred to as a heap, though the larger and less defined the pile, the more likely it is to be thought of as a heap. One can have a heap of sand or sawdust, but not a pile, which would suggest some magical work of magnetism or levitation. For something to be orderly in construction, the units of which it is made up need to be visible and divisible. Size here is therefore to be correlated with countability. A pile is broadly equivalent in British English to a bunch, which equally will tend to consist of a number of distinguishable elements – as in a bunch of flowers, keys or coconuts. In American English, by contrast, 'bunch' seems to have lost the necessary suggestion of numerable plurality: 'a bunch of stuff' sounds odder to British ears than to American. When Francis Galton and others set about measuring the number of shocks that a railway passenger could expect to receive during the course of a journey, the word 'shock' might just have retained a little of its original meaning of a sheaf (as in 'a shock of red hair'). The individuated shocks of the train on the rail may have seemed to have something in them of the singular-multiple. A 'lump' stands in the same relation to a 'mass' as 'pile' does to 'heap', that is, it seems more compact and choate. A 'mass' of inflamed tissue may be relatively small, smaller than a lump under some circumstances, but a mass nevertheless always suggests a vaguer, less determinate kind of magnitude. This is why a knot of people is smaller than a mass of people, for a knot has a defined form, with a clear edge between its inside and its outside, while a mass suggests not just something whose volume is undefined, but something that is in the process of amassing, or is like to engorge. Masses are not just massive, but massing.

Language seems particularly good at making units out of aggregated multiplicities. In such concepts, in which language abounds, words and numbers intersect and interact in surprisingly intricate ways. One may see this typified in the expression 'a certain number'. The number here cannot be referred to exactly, because the actual or precise number is not known, but it can be acted on and transacted with. It is certain that some particular number or other must be implied, if not specifically meant, by 'a certain number', even though that number may never emerge. The 'certain number' of which a group, a crowd, a swarm, a mass, a multitude are (is) made is certainly some number, though it is uncertain which.

The class of ancient philosophical conundrums known as sorites paradoxes trade on the fact that in aggregates like heaps, certainty and uncertainty are compounded. In the most famous sorites paradox, propounded by Eubulides of Miletus, the following chain of reasoning is offered. If a heap of grain consists in a large but indeterminate number of grains, the removal of one grain can never be enough to cause the heap to be anything less or other than a heap. We may in fact say that it is part of the definition of a heap, as a roughly defined aggregate, that it should have this characteristic. Something to which the removal of a single grain would make such a difference could not plausibly be described as a heap. So, to be essentially a heap, a heap must be only more-or-less a heap. And yet, if one grain is repeatedly removed, there will in fact come a point – the point at which there is only a handful of grains left, or perhaps only one, or perhaps even none at all – when the heap will in fact have ceased to be recognizable as a heap. Since it is the removal of the grains one by one that makes the difference, there ought to be a point at which the heap ceases to be a heap – yet it is impossible to specify quite what that point is, if it is really true that the removal of one grain can never make that much difference. The paradox can be run the other way round. If adding one grain will never suffice to make something a heap, then there is no point at which 1 + 1 + 1 + 1 . . . individual grains will ever definitively tip over into being a heap. This is the form of the paradox to which Samuel Beckett alludes in *Endgame*: 'Moment upon moment, pattering down like the millet grains of . . . that old Greek. And all life long you wait for that to amount to a life.'[1]

These questions bear on the force in language and thought of what I will call the imagination of multitude. Multitude must always in fact be imaginary. In the Old Testament, multitude often seems to carry the intimation of unlimited bounty; multitudinousness is a promise. The word first appears in Genesis, in the promise made to Hagar, who has fled from her mistress Sara, bearing Abraham's child: 'And the angel of the LORD said unto her, I will multiply thy seed exceedingly, that it shall not be numbered for multitude' (Genesis 16:10). Jacob alleges of the Lord that he has said 'I will surely do thee good, and make thy seed as the sand of the sea, which cannot be numbered for multitude' (32:12). In Deuteronomy, the Lord says 'the LORD your God hath multiplied you, and, behold, ye are this day as the stars of heaven for multitude' (Deuteronomy 1:10), and the formula 'as the stars of heaven' is repeated twice more in the book. In Judges, we read that 'the Midianites and the Amalekites and all the children of the east lay along in the valley like grasshoppers for multitude; and their camels were without number, as the sand by the sea side for multitude' (Judges 7:12). The phrase 'it shall not be numbered for multitude' is a pleasing contrivance. We could say it means: the number of items is so large that one cannot attain to it through numbering. Multitude of this kind is not at all outside the order of number; in fact it is its numerousness and nothing but its numerousness, the fact that there are so many numbers on the way to its number, that stands in the way of assigning it its correct number.

The word 'multitude' seems to undergo a change of magnitude between the Old and New Testaments, as it moves from the cosmos to the polis. In the New Testament, the multitude is often a crowd – 'the whole multitude sought to touch him' (Luke 6:19); 'an innumerable multitude of people, insomuch that they trode one upon another'

(Luke 12:1); even the disciples are a 'whole multitude' (Luke 19:37). Multitude is no longer exorbitant on a cosmic scale. Even in Revelation, the 'great multitude, which no man could number' is 'of all nations, and kindreds, and people, and tongues' (Revelation 7:9), 'great multitude, which no man could number' translating 'ὄχλος πολύς ὃν ἀριθμῆσαι αὐτὸν οὐδεὶς ἐδύνατο'. In Revelation 19:6, there is a more impersonal 'voice of a great multitude [again, ὄχλου πολλοῦ], and as the voice of many waters, and as the voice of mighty thunderings', but the fact that this is a voice that nevertheless coheres into an utterance – 'saying, Alleluia: for the Lord God omnipotent reigneth' – reins in its grandeur.

To imagine a multitude is to imagine something not fully imaginable, a singularity formed of pure multiplicity. A multitude is a finite indefinite, precisely because it must be a multiplicity of some distinguishable and reckonable kind of thing. We do not usually mean by a multitude a pure multiplicity, of huge numbers of different kinds of thing. A multitude both is and is not countable. It is a plurality of things that goes beyond counting, but nevertheless must consist of things that are in principle countable, things that can be counted as ones of something – loaves, fishes, stars, grains of sand. A multitude is a piece of implicit or deferred numeration, not that which is beyond number, or outside the order of number, but a kind of innumerable number, like an unspeakable word. There are many kinds and degrees of multitude. We might even say that there is an infinity of different multitudes, though multitudes themselves are not and cannot be infinite, because they must add up to some unknown quantity, some countable, but uncounted number. A multitude is a kind of *x*, an algebraic placeholder.

As such, the word 'multitude' has an implicit reference to the intersection of words and numbers. It is a word standing in for an unspecified, unspoken number, or even for a particular, subjunctive feeling relating to that sense of a number. One example of the spectrum of feelings that may be associated with the idea of multitude may be what W. B. Yeats describes as 'the emotion of multitude'. In an essay of that name of 1903, Yeats laments its absence from the 'clear and logical construction' of French drama, contrasting it with Greek drama, which

> has got the emotion of multitude from its chorus, which called up famous sorrows, long-leaguered Troy, much-enduring Odysseus, and all the gods and heroes to witness, as it were, some well-ordered fable, some action separated but for this from all but itself.[2]

Often, this feeling is provoked by the recognition that 'the sub-plot is the main plot working itself out in more ordinary men and women, and so doubly calling up before us the image of multitude'.[3] Yeats's phrase 'emotion of multitude' is more exact than it may at first appear. It would be easy to see Yeats as writing simply in praise of Celtic blur here, of 'vague symbols that set the mind wandering from idea to idea, emotion to emotion'.[4] But it is not just the slide from emotion to emotion that Yeats refers to but a more exact and emphatic singular: the emotion of multitude, emotion centred on and responsive to the fact of multitude itself.

Immanuel Kant's reflections on the mathematical sublime provide us with a way of considering this interference of the numbered and the unnumbered. Kant distinguishes the capacity of reason, as exemplified in mathematical thinking, to think to infinity and, so to speak, beyond – this he calls 'apprehension' (*Auffassung*) – from the aesthetic capacity of the mind to grasp the magnitude as a whole, or a *Zusammenfassung* (comprehension):

> Apprehension presents no difficulty: for this process can be carried on *ad infinitum*; but with the advance of apprehension, comprehension becomes more difficult at every step and soon attains its maximum, and this is the aesthetically greatest fundamental measure for the estimation of magnitude.[5]

Kant asks us to think of a spectator of a pyramid or of St Peter's in Rome who is standing too close to take in the whole at a glance. Kant says that this produces first a kind of striving to overcome the limit on the synthesizing imagination, and then a kind of disappointed recoil – which nevertheless produces a kind of satisfaction:

> here a feeling comes home to him of the inadequacy of his imagination for presenting the idea of a whole within which that imagination attains its maximum, and, in its fruitless efforts to extend this limit, recoils upon itself, but in so doing succumbs to an emotional delight.[6]

Thus it is that, as Kant puts it, 'that same inability on the part of our faculty for the estimation of the magnitude of things of the world of the senses to attain to the idea, is the awakening of a feeling of a supersensible faculty within us.'[7] As Andrzej Warminski describes it, this turns an incapacity into a paradoxical kind of 'faculty'. We may say that the imagination of multitude encompasses a similar movement of the mind, that somehow captures and folds back on itself the open capacity of mathematical reasoning.

Kant here seems to complicate fatefully the distinction on which his analysis depends, between the mathematical and the non-mathematical. For the satisfaction in disappointment involves a kind of contusion of his structure, in which the mathematical is used to form the non-mathematical. The imagination of multitude may be a result of this same contusion. Kant notices that our powers of relative observation are greatly increased by various technical devices:

> nothing can be given in nature, no matter how great we may judge it to be, which, regarded in some other relation, may not be degraded to the level of the infinitely little, and nothing so small which in comparison with some still smaller standard may not for our imagination be enlarged to the greatness of a world. Telescopes have put within our reach an abundance of material to go upon in making the first observation, and microscopes the same in making the second.[8]

This presumably means that the sense of the gap between the open relativity of apprehension (*Auffassung*) and the closed absoluteness of comprehension (*Zusammenfassung*) increases. The contemporary imagination of multitude colonizes this gap between scales, which is opened up in many more places by the grasp-exceeding reach of contemporary instances of multitude.

This play between number and the innumerable operates with particular force in the Bible, since its texts, and especially the books of the Old Testament, are so thronged with number and numeration – ages, heights, weights and measures of all kinds, counted out in what can seem like obsessive detail. If a word like 'multitude', and the various words it translates, seems to move language beyond number, or to use language to move number beyond itself, this is against a background of careful and systematic coordination of word and number. If it is characteristic of a sacred text that it evokes the ineffable and the uncountable, it can also be characterized by a kind of sacred exactitude and totality. This is encouraged by the Kabbalistic tradition of *gematria* (probably derived from Greek *geometria*), which assigns numerical values to letters and words and reads significance into the quantitative rhymes between words and names. The Masoretic text of the Hebrew Bible derives from the work of the scribes and scholars working between the seventh and eleventh centuries, who became known for the accuracy of their techniques. It became common for them to add marginal notes, known as *masorah*, which deal with errors and variants, and give details of vocalization and accentuation. But one of the most important means of checking the accuracy of texts and stabilizing their transmission was through counting of letters and the calculation of word-use statistics.

Counting procedures are important in many religions, but perhaps particularly so in Judaism, for example in the observance known as Sefirat HaOmer or Counting of the Omer, which involves the counting of the 49 days between the sacrifice of an omer-measure of barley at Passover and the offering of wheat brought to the temple at Shavuot, as required by Leviticus:

> 15. And ye shall count unto you from the morrow after the day of rest, from the day that ye brought the omer of the waving; seven weeks shall there be complete;
> 16. even unto the morrow after the seventh week shall ye number fifty days; and ye shall present a new meal-offering unto the LORD. (Leviticus 23:15–16)

If God is beyond measure and numeration, then the apprehension of God is often regarded as a matter of counting. In Kabbalah, the En Sof, or infinite principle, reveals itself through ten attributes or emanations, known as the Sephirot, the plural of Sephirah, meaning counting. These principles are both themselves numbered and are number, or plurality itself.

Counting out has its complement in the practice of number divination. Where counting aims to get language and bodily action to march precisely in step with the operations of number, number divination exploits the opposite principle, of the blindness to number, and the capacity of multitude to dazzle or distract. Sieve holes seem to have a related function in Korean folklore, in which goblins were thought to be kept at bay with a sieve hung on a gatepost:

> People believed that goblins descended and started counting the tiny holes in the sieve, got confused while counting, and were forced to recount over and over. When the goblins heard the roosters crow, they stopped counting the holes, complained about the imminent sunrise and hurried back to their abodes in the skies.[9]

Here the multiplicity of the sieve holes captures or detains the goblins through the compulsion to count, combined with the dizzying difficulty of keeping count. It is common in divination practices for there to be a surrender to a kind of counting procedure that itself muddles or obscures the count. One interesting piece of seventeenth-century folklore seems also to draw on the hypnotic powers of the sieve. In his *Entertainment at Althorp* (1603), Ben Jonson says that the fairy Queen Mab 'Trains forth midwives in their slumber/ With a sieve the holes to number'; in *Leviathan* (1651), Thomas Hobbes refers to 'counting holes in a sive' as one of the 'Prognostiques of time to come' and Richard Levin has persuasively interpreted this as some kind of counting procedure: 'Probably one counted the holes while reciting some formula (which may have named the alternative possibilities) and the answer was determined by the number of holes, or, more likely, by the end of the count.'[10]

Counting procedures like eeny-meeny-miny-mo depend on a similar principle of distraction, in which the words, while seeming themselves to form a count, in fact work to obscure it, succession concealing summation. A similar work of distraction is also at the heart of a riddling rhyme like 'As I was going to St Ives'. The very name for such playful procedures – riddles – gives a clue to the role of the uncountable multiplicity in them. Such riddles often have at their heart plays between nothing and number, for example in John Lennon's folkish lyric from 'A Day in the Life', which evokes the task of counting the four thousand holes in Blackburn, and the eventual satisfaction of knowing how many holes are required to fill the Albert Hall.

Divination procedures work equally well with different kinds of multitude – flights of birds, entrails, stars, tea leaves. What seems to matter is that some procedure is employed to make a seemingly unordered mass reveal an order – typically by making or enabling a decision, for example a simple yes or no to some enquiry. Books such as the Bible or the works of Virgil are often used for the process of what is known as a sortilege, literally *legere*, the reading, of a *sors*, a lot, or portion. It seems to be important that the book in question is not only regarded as sacred but is large enough to be thought to contain the whole world in possibility. With the growing importance of the Torah, the book would increasingly be consulted in preference to prophets. The principle is articulated by Pieter van der Horst:

> Since all that God had, has, and will have to say to mankind is contained in the Torah, and since he can be trusted to guide and control this process of consultation, the answer is incontrovertible, in fact a prophecy (*nevu 'ah*). As one of the early rabbis (Ben Bag-Bag) is reported to have said about the Torah: 'Turn it, and turn it again [i.e. study it from every angle] for *everything is in it*' (Mishna, *Avoth* v 22), not only everything of the past, but also of the present and of the future.[11]

Rabbis would ask children what verses they had studied that day in school, and to take the answers as good or bad omens.[12]

Christians also used the Bible as a divinatory resource, though this practice was often condemned,[13] and the works of Homer and Virgil, whose status was almost as high in the classical world as that of the Bible among the Jews, had also been used for bibliomancy, the latter by means of the *Sortes vergilianae*.

Chance was sometimes combined with the operations of particularly significant numbers, as, for example, in the practice recorded among Lithuanian rabbis of opening the Hebrew Bible at random, counting seven pages, then reading seven lines down, with the resulting verse taken to be the revelation.[14] There were secular versions of this kind of magical intersection of word and number, for example the *Sortes astrampsychi*, a collection from the second or third century CE of some 92 questions and 1,030 answers said to be the work of the mythical magician Astrampsychos. A complex numerical procedure governs the relationship between question and answer:

> The enquirer first looks in the list of 92 numbered questions to find his question or the one most like the question he wants to raise. Then he chooses by some kind of sortition or selects in his mind a number between 1 and 10 and adds it to the number of his question. The sum thus reached has now to be looked up in a list of oracular gods with a concordance following after the list of questions. The concordance indicates by means of a number after the god's name the 'decade,' i.e., the section with ten possible answers. In that decade the answer is found under the number that was chosen by lot. For example, your question is, 'Will I get the woman I want to have?' This is question no. 29. You draw by lot or select the number 7, so the total is 36. In the list of oracular gods you find under 36 Hephaestus, and after his name the concordance number 27. Decade 27 has under number 7 the following answer to your question: 'Yes, you will get the woman you want, but much to your detriment!'[15]

Bibliomancy seems to depend on another version of the alternation between number and imagination discussed by Kant. The use of chance depends on and is itself the proof of an open world of possibilities that is too large for the mind or imagination to encompass. But a sacred text is held to include all those possibilities, in a set of one-to-one pairings between text and world, which are ultimately the proof of the absolute two-in-one mirroring of the world in the mind of God:

> The universe is one close-meshed unit; heaven and earth, animals, plants, angels, demons, man, all are creatures of God, manifestations of His will, all so sensitively intertwined that each reacts immediately to the slightest alteration in the composition of the whole . . . Events predetermined in the mind of God impinge upon one or another aspect of His universe long before they reach the final stage of occurrence on earth; the superior sensitivity of certain parts of the world, and even of parts of man's immediate environment and body, makes them responsive to what is yet to be long before it is.[16]

As Michel Foucault writes, in such a world, 'the universe was folded in upon itself: the earth echoing the sky, faces seeing themselves reflected in the stars, and plants holding within their stems the secrets that were of use to man.'[17] Divinatory practice, like many other magical procedures with language, treats words as quasi-numerical, as part of a vast system of one-to-one correspondences, a system of twos that guarantees a universe of clearly distinguished yet coupled ones. There will always be more than the human mind can comprehend, yet certain procedures will allow the breaking through of knowledge of some portion of this otherwise unimaginable totality.

The word *lot*, and its plural *lots*, are subject to a similar semantic movement from singularity to generality, from the specific to the unspecified. A lot, perhaps from a Germanic word for a piece of wood used for casting lots, is a particular allocation, that which is allotted, whether by human process or by fortune – 'my lot in life'. The Old English *hlot* is in fact used to render Latin *sors*, a portion. But, as the word *lot* came to be used like the word *sort*, to signify a number of things or persons of the same kind, it seemed to move in the direction of the indefinite. The word also begins in the process to gather a sense of deprecation, as in a phrase like 'that lot', or 'you lot', and with a drift towards the word *load* – a 'bad lot', 'a load of rubbish'. 'The lot' comes to mean 'the whole measure', usually evoked in a way that suggests an unparticularized multitude. The OED citations for 'the lot' typically begin with a series of items that is then broken off, sometimes incorporating a dash or series of dots to signify indefinite extension, 'and so on': 'It was to be a big wedding – the full treatment – Royalty – the lot'; 'They are said to cure everything from rheumatism to ringworm, colic to snake-bite . . . – the lot'; 'The death of his father . . . triggers off a crisis for him too, producing a temporary breakdown, dismissal from his job, separation from his wife, the lot'; 'They've searched the island twice – helicopters, dogs, the lot.'

The authority of the idea of definite but unspecified quantities hidden in uncountable masses may relate closely to the taboo on counting that is found in many cultures.[18] In Africa in particular, but many other places besides, the taboo on counting embodies the fear that making an exact count of living beings or possessions, children, cattle or crops, will put them in danger of destruction: not counting your chickens before they are hatched may be a mild form of this inhibition.[19] The taboo is often avoided by means of matching or tallying procedures, such as reciting a verse or eeny-meeny-miny-mo formula, which allow particular quantities to be established without resort to counting.[20] Indeed, Abraham Seidenberg suggests that the use of fingers and toes may be important, not as the origin of base-10 counting, but rather as a means of circumventing the taboo on counting out loud.[21]

The influence of the King James Bible has probably helped to ensure that the word 'multitude' belongs to a distinctly religious register, evoking awe and the sense of sublimity. But the powerfully religious force of the word has lately passed across into political discourse through the adoption of the term 'multitude' by political theorists Michael Hardt and Antonio Negri, who urge its substitution for more monolithic terms like 'mass' or 'people':

> The people is one. The population, of course, is composed of numerous different individuals and classes, but the people synthesizes or reduces these social differences into

> one identity. The multitude, by contrast, is not unified but remains plural and multiple. This is why, according to the dominance tradition of political philosophy, the people can rule as a sovereign power and the multitude cannot. The multitude is composed of a set of *singularities* – and by singularity here we mean a social subject whose difference cannot be reduced to sameness, a difference that remains different . . . the challenge posed by the concept of multitude is for a social multiplicity to manage to communicate and act in common while remaining internally different.[22]

What holds the multitude together is the fact of their shared opposition to the order of capital: 'Our initial approach is to conceive the multitude as all those who work under the rule of capital and thus potentially as the class of those who refuse the rule of capital' (*Multitude*, 106). This is therefore a kind of subjunctive multitude, the multitude that is immanent in the mere multiplicity, a multitude-to-come that would actualize the creative force of the common interests of those who refuse the rule of capital: 'Today we create as active singularities, cooperating in the networks of the multitude, that is, in the common' (*Multitude*, 135).

An impatiently positivist view might be that this displays a fatal uncertainty at the heart of Hardt and Negri's project. Sometimes the value and promise of multitude lies in the creative labour its constituents have and hold as a kind of commonwealth and the common aim they have of surpassing or circumventing the rule of capital. At other times, the power of multitude lies simply in the fact of its/their very multiplicity, as the lack of commonness. The power of multiplicity is therefore dangerously demonic, insofar as the demonic just means the multiple. Hardt and Negri relate the power of the multitude to Dostoevsky's *The Devils*:

> *What is so fearsome about the multitude is its indefinite number, at the same time many and one. If there were only one conspiracy against the old social order, like Dostoyevsky imagines, then it could be known, is confronted, and defeated. Or if there were many separate, isolated social threats, they too could be managed. The multitude,*

> *however, is legion; it is composed of innumerable elements that remain different, one from the other, and yet communicate, collaborate, and act in common. Now that is really demonic!* (Multitude, 140)

Though understandable, impatience at the apparent ambivalence of the idea of multitude would perhaps miss the point. Multitude as demonic commons is an attempt to articulate a special kind of number, or a special way of articulating number. Perhaps the number of multitude is not to be rendered as a single, simple integer, but as a kind of oscillating number-function that moves, only semi-predictably, between the polarities of the one and the many. Perhaps, indeed, this is the essential plurality of the idea of multitude – not the simple plurality of that which, always being more than one, is therefore always less than One, but the refusal to settle into either the singularity of oneness or the singularity of plurality.

What makes the multiplicity of different kinds of classes cohere is not any principle of organization arising from within itself, but an external fact that makes mere multiplicity genitive, that is, a multiplicity *of* certain items or instances. This heteronomous principle is capital, which, here, as almost everywhere else nowadays, is the great, all-explaining 'Count-As-One' of the contemporary political imagination. What allegedly unites the forms of the multitude, what brings its instances into commensurability and constitutes its historical horizon of common purpose, is capitalism. And what is the core, cohering principle of this capitalism but the subjection of the incommensurable to measure for the purpose of extracting profit? The underlying principle of counted-as-one capitalism, the very principle that allows it to be counted as one, and therefore in turn permits the otherwise merely teeming multitude of its resistances also to be counted as one, is nothing but the principle of counting itself.

All this is both product and producer of fantasy. Capitalism used to be one thing among many, a particular set of economic relations that could be reliably distinguished from other actually existing or possible economic relations. Now, the function of the idea of capitalism is to act as the horizon of horizons; no longer one thing among many, it is the oneness of the many. In fact the only capitalism there is – the only thing that makes 'capitalism' the great count-as-one that it increasingly has to be – is that which is reciprocally implied in the multitude that itself can only be unified by the fabled object of its resistances. Capitalism depends on the multitude that depends on it. Capitalism and multitude are reciprocating genitives. The multitude is the multiplicity of resistances to capital; capital is what the multiplicity of these resistances converges on. There can be no multitude without count-as-one capitalism, but there can be no capitalism that could count as one unless it gave rise to a multitude that can be counted as one, can be grasped as more (that is, less) than a mere multiplicity.

LEGION

Religion is what binds – *ligare*. And this binding often involves the imagination of number. Monotheism has been described by Peter Sloterdijk as deriving from an allergy to the number two, which we might see as the fundamental and formative fracture within the order of number itself, and striving to bring 'everything down to the number one, which tolerates no one and nothing but itself'.[23] Number is useless without the capacity to count, that is, to take account of pluralities. But plurality, the fact that there can be many different kinds of singular, many different things that may be counted as one, is always a threat to the establishment of number. Monotheism depends on the claim that there is no god but God, but there are at least three kinds of monotheism of which account must be taken; in one of the three, the Christianity formed around the Trinity, the one *is* three. Monotheism is haunted by set-theoretical paradox, because there must always be an excess, of outsiders who cannot be accommodated to the Kingdom of Heaven. It can only be one if it is not two, but if it is only one and not two, then this means it does not in fact include two, which means it cannot really be one. Oneness requires there to be some surplus or remainder, which will nevertheless then compromise any claim of any One to be All-in-one. This remainder has many names: evil; the Devil; sin; woman; time. The singular-plural, or singural name for this manyness is 'Legion', the name we give to the one-who-is-not-one, who does not have a single name. Hardt and Negri say that the self-designation by the Gerasene demoniac spoken of in the Gospels embodies a '*metaphysical threat*', for, '*since it is at once singular and plural, it destroys numerical distinction itself*' (*Multitude*, 138). So the demoniac is, like the Golem, another (one in a numerical series, in fact) avatar of multitude itself. Hardt and Negri would like multitude to put number in jeopardy because number as such is taken to be of the beast. Just as economic exploitation requires the brutal violation of living labour by number, so

> *Political thought since the time of the ancients has been based on the distinctions between the one, the few, and the many. The demonic multitude violates all such numerical distinctions. It is both one and many. The indefinite number of the multitude threatens all these principles of order. Such trickery is the devil's work.* (*Multitude*, 138–9)

Demons are one of a number of beings-of-number, beings whose ontology consists of their numerousness, that recur in the thinking of capitalism and its opponents. Zombies have also often been seen as metaphors for the half-dead enslaved proletariat of capitalism, but, in some recent formulations, it is capitalism itself which is zombie-like, because it is a kind of death-in-life, a virus that, having no living creative force of its own, parasitically steals the living labour of others, and transforms it into the pseudo-life of number. So, we read, '21st-century capitalism as a whole is a zombie system, seemingly dead when it comes to achieving human goals and responding to human feelings, but capable of sudden spurts of activity that cause chaos all around.'[24] Jacques Derrida is among those who have pointed to the Gothic streaks in Marx's writing, which are tied to the desire to assert the principles of life against death, or, worse, the counterfeiting of life:

> Marx does not like ghosts any more than his adversaries do. He does not want to believe in them. But he thinks of nothing else . . . He believes he can oppose them, like life to death, like vain appearances of the simulacrum to real presence.[25]

Zombies embody the horror of number, the horror of the fact of there being, not just larger numbers of everything, but ever larger numbers of numbers. Zombies embody the condition of all numbers, which is to impersonate life, since to be a mere number is not really to be alive at all. But there is also a strange fecundity in zombies, who transmit the contagion of their immortality seemingly without constraint. In fact, zombies share their quality of multitudinousness with demons. The connection with the cosmic multitudes of the Old Testament is that their exponentiality, their number-outnumbering-itself, embodies an essentially economic promise, of a kind of cosmic compound interest. They have an exorbitant multitude which oscillates between bounteousness and abomination. Angels and demons are both entities of pure number; though they are nameable and countable, angels in particular being carefully ranked, into seraphim, cherubim, principalities, powers, archangels and so on, their essence seems to consist in a pure capacity for prodigious expansion, beyond naming or numbering.

Religion, like politics, binds the many into one, often by means of the idea of a collective body. For Hardt and Negri, the body of multitude is itself multiple, not just because it is made up of multiplicity, but in that it comes into being between two possible forms of imaginary incorporation. In the first such incorporation, multitude is the body-double of capital. What Hart and Negri call, extraordinarily, 'the real flesh of postmodern production' is 'the object from which collective capital tries to make the body of its global development. capital wants to make the multitude, wants to make it into a people' (*Multitude*, 101). But this false or forced incorporation is the very means by which multitude may find its real incarnation:

> This is where, through the struggles of labor, the real productive biopolitical figure of the multitude begins to emerge. When the flesh of the multitude is imprisoned and transformed into the body of global capital, it finds itself both within and against the processes of capitalist globalization. The biopolitical production of the multitude, however, tends to mobilize what it shares in common against the imperial power of global capital. In time, developing its productive figure based on the common, the multitude can move

> through Empire and come out the other side, to express itself autonomously and rule itself. (*Multitude*, 101)

It is scarcely surprising that the Christian notion of the mystical body of the Church is invoked in section 2.2 of *Multitude*, entitled 'De Corpore'. Hardt and Negri look to the possibility 'that these common singularities organize themselves autonomously through a kind of "power of the flesh" in line with the long philosophical tradition that stretches back at least to the apostle Paul of Tarsus' (*Multitude*, 159). That such a view of multitude forms an exact parallel to what is commonly and contemptuously called 'corporate culture' is presumably perfectly apparent to Hardt and Negri. Indeed, one of the signs of the mixed body of which the idea of the multitude is made up is the fact that it has attracted explication as theological politics, for example in the comparisons of the decentred multitudes of the Occupy movement to the *ochlos*, the word used frequently in the Gospels for the crowd that attended Christ.[26] The Gospels do not speak entirely consistently about this, however, for Luke seems to use the word *ochlos* in its more derogatory sense as 'mob' (Citron 1954, 410), and generally has a less benign view of multitude; indeed, at one point, he identifies multitude with madness, saying that those 'vexed with unclean spirits' are ἐνοχλούμενοι ἀπὸ πνευμάτων ἀκαθάρτων (Luke 6:18), *ochlos* providing the verb for the demonic crowding out of their senses.[27] Western conceptions of multitude come together with Eastern in the Korean notion of *minjung*, the people.[28]

To be sure, there is much incoherence in Hardt and Negri's notion of multitude, but it is a telling and expressive and possibly valuable incoherence, if only because of their agonized attempt to assign a value to incoherence itself, in the one that comes to be one in refusing to join up as one. Their work may be regarded as an instance of the thinking of the multiple in itself called for by Michel Serres at the beginning of his meditation on beginnings, *Genesis*. We find it hard not to either break aggregates down into their atomic constituents or round them up into unities:

> We want a principle, a system, an integration, and we want elements, atoms, numbers. We want them, and we make

> them. A single God, an identifiable individual. The aggregate as such is not a well-formed object; it seems irrational to us. The arithmetic of whole numbers remains a secret foundation of our understanding; we're all Pythagoreans. We think only in monadologies.[29]

Serres makes out a 'new object for philosophy',[30] in a pure, which is really to say impure, multiplicity:

> The multiple as such. Here's a set undefined by elements or boundaries. Locally, it is not individuated; globally it is not summed up. So it's neither a flock, nor a school, nor a heap, nor a swarm, nor a herd, nor a pack. It is not an aggregate; it is not discrete. It's a bit viscous perhaps. A lake under the mist, the sea, a white plain, background noise, the murmur of a crowd, time.[31]

It was often reported during the nineteenth century that the most primitive peoples – by which was often meant indigenous Australians – had only three counting numbers: '1. Wagul. 2. Boola. 3. Brewy. When a number exceeds *three*, they use the phrase *murray loolo*, which signifies an indefinite number.'[32] There will always be an area of inexactitude, which is not beyond number as such but is beyond exact number, in any apprehension of number. It is a sign of childishness for us nowadays for somebody to imagine that there might be a biggest number of all, since we know in principle that, no sooner have we invented a term for a very large number – a billion, a light year or a terabyte – than that number can be multiplied or squared and so on, to yield an even larger number. But we are not so far removed historically from periods at which it was still possible to marvel simply at the existence of vast numbers – for example, in the early seventeenth century, at the realization of the huge number of permutations required to 'ring the changes' completely of a set of bells. We may perhaps call this the Archimedean point, after the story that Archimedes asked for payment in the form of a grain of wheat placed on a square of a chessboard, two grains on the next, four on the next, and so on, doubling each time – the total exceeding the annual grain production of Egypt. The Archimedean point is fulcrum at which, by extrapolation, one seems able to reach imaginatively out into the realm of pure quantity, allowing one to inhabit an island of numerative capacity in the midst of an ocean of pure number. Kant's mathematical sublime may describe a similar fulcrum or moment.

We are required to extend ourselves, not just mathematically, but also imaginatively, further and further out into this indifferently grey and churning ocean of pure number. As a result the dim void becomes differentially textured and striated, becoming a kind of cartography of significances and sensitivities. We are accustomed to think, and we are certainly everywhere assured and instructed, that art and literature are concerned with the exploration and preservation of the kind of 'human' qualities that are under threat from the 'inhuman' dominion of number. But only the numbest kind of abstract thinking, which might itself be taken as an instance of the kind of calculative rationality of which it dreams, allows us to imagine that qualities and quantities can be separated in this way. What we call qualities – the pain of a toothache, the pinkness of a sunrise – are almost always quantical; always, that is to say, a matter of more or less consciously calculated degree. And the quantical is always itself a matter of differential qualities. Despite the conventional revulsion from number, a revulsion which Hardt and Negri share even with mathematical writers like Alain Badiou, their work forces a recognition that all responsible and intelligent thought must now pass through number, and in ever more complex ways. We do not need to have the synaesthete's conviction of the blueness of fives or the pointiness of primes to be able to apprehend and appreciate these qualities. Far from being at threat from or themselves opposed to the realm of number, art and literature (and not, of course, just these, but also pushpin and pop music) are indispensable agents in the 'existing' of number, to borrow a term from Jean-Paul Sartre, the embedding of numerical awareness and sensitivity into more and more areas of social and personal life.[33] Nowhere is this more the case than in the different ways in which the imagination of multitude is elaborated.

6

HILARIOUS ARITHMETIC

WORKING OUT

To say that mathematics is something to be done, as I observed at the beginning of Chapter Two, is to recognize its deserved reputation for being hard work. Mathematics has to be done, to be worked out, in a way that other intellectual operations do not. If children are enjoined to 'show their workings' in mathematics, that is because mathematics, unlike music, say, or geography, consists in its workings, rather than its outcomes. If the world is indeed written in the language of mathematics, there is labour in that deciphering. And yet there is also no mental discipline which seems more to exemplify Michel Serres' principle that in fact all work amounts to sorting, whether hard, through the physical movement or transformation of things through the expenditure of physical energy, or soft, through the sifting of information or ideas.[1]

The work involved in mathematical procedure is between the soft and the hard. There is always some kind of cost in computational effort involved in the forming of every calculation, whether that calculation be performed by a supercomputer or a sulky second-former. The idea of infinity means that one can carry on adding one to any number without ever coming to an end. But Brian Rotman has suggested that there is a cost even to the elementary action of counting, and indeed, makes it one of the reasons why he says we must abandon belief in the existence of infinite quantities; for there must come a point at which the computing resources necessary simply to keep in mind the largest number ever articulated, and then add one to it, would exhaust all the energy in the universe. There would come a point at which one would be bound to lose count, whatever system were devised for keeping it.[2] Perhaps this is why the Godhead is sometimes identified, not just with the infinite, but with the infinite capacity to keep count: even the very hairs of your head are all numbered, Christ assures us, on an occasion 'when there were gathered together an innumerable multitude of people, insomuch that they trode one upon another' (Luke 12:7, 12:1). The uneasiness of certain groups of believers about the difficulties for God of reassembling bodies that have been dissolved by fire, as in cremation, as compared to bodies that have been kept (as they imagine) relatively intact, as in burial, is the hint of an impious doubt even among the most devout of the operational limits even of the Good Lord's molecular database.

But it feels as though one need not in fact resort to this kind of operation, which, tellingly perhaps, mathematicians call a 'brute force' operation. It feels as though applying the logical principle that there must always be a one that can be added exacts no cost at all, any more than simply and immediately seeing that $2 + 2 = 4$ does. I can prove that $2 + 2 = 4$ by counting, but mathematics means not having to count, even if its results depend upon the possibility of this application of brute force in the first or final instance.

Indeed, mathematics operates between these two polarities, of the huge and the minimal, the massive and the negligible, the energetic and the angelic, the quantical and the nonquantical. Indeed, this is the reason why so many scientists, somewhat to the surprise of those in the humanities, insist that mathematics cannot be scientific. For mathematics depends upon – really, in fact, consists in – the generation of proofs, which go (almost) to infinite pains to show that the complex equivalences of quantities and relations made out by calculation have existed all along. The largest known prime number, discovered on 7 January 2016 by Curtis Cooper at the University of Central Missouri, is $2^{74,207,281}-1$. It took 31 days of continuous computing to prove the primality of this number. And yet that considerable outlay of conjoined human and mechanical work yields no outcome that makes any difference to the way things are and always must have been in the world of numbers. The largest prime happens to be one of only 49 known Mersenne primes, that is a prime number formed from 2^n-1, where n is itself a prime number. Like all prime numbers, it has been there all along, and so might perfectly well have been stumbled upon by accident, rather than as a result of the assiduous searches being conducted worldwide by the Great Internet Mersenne Prime Search (GIMPS).

One of the links between comedy and mathematics depends on this strange identity of exertion and ease, of almost everything and scarcely anything. One might even say that there is a kind of drawn-out absurdity in the procedure of solving, the effort to show that something is exactly what it was all along. Mathematical proof depends upon demonstrating forms of equation that depend upon the increasingly radical non-equality of the effort of the proof and its outcome.

If mathematical proofs can be thought of as the solving of puzzles, they can, by the same token, and using almost the same terms, also be seen as having the structure of a joke, according to the terms of the well-known relief theory of comedy. This receives a formulation in the work of Immanuel Kant, who writes in his *Critique of Judgement* that '*Laughter is an affect arising from a strained expectation being suddenly reduced to nothing*.'[3] He gives the example of the following joke, or 'joke':

> an Indian at an Englishman's table in Surat saw a bottle of ale opened, and all the beer turned into froth and flowing out. The repeated exclamations of the Indian showed his great astonishment. 'Well, what is so wonderful in that?' asked the Englishman. 'Oh, I'm not surprised myself,' said the Indian, 'at its getting out, but at how you ever managed to get it all in.'[4]

The laughter, such as it is, prompted by this story comes about, says Kant, because 'the bubble of our expectation was extended to the full and suddenly burst into nothing.'[5] There is an interesting wrinkle in this particular example, since the process whereby something comes to nothing in the response to the joke is mirrored by the terms of the joke itself, which is itself precisely about something turning into something that is as good as nothing (froth). It is as though the joke were showing its own workings, which is no doubt the reason for its utility for Kant, even though he does not mention it.

An expectation is created, then dissipated: a something is suddenly transformed into a nothing. We might say that the simple formula for this operation would be $1 - 1 = 0$. Subtract something from itself, and the result will be nothing. But the joke as explicated by Kant seems also to accord with another formula, according to which something is shown to be equivalent to nothing, something is shown to have been nothing at all, all along. The formula for this would be $1 = 0$. The issue touched on here, or, as one is tempted to say, in view of the explosive implications, touched off, is the complex one of whether zero is in fact to be regarded as a number at all. In many instances of comedy, zero is not so much a particular quantity as the sudden abeyance of the quantitative as such. Zero is not so much a position on the number line as a gap in it, or an intersection of that number line by nonnumericality. Viewed in this way, zero would not be in the same plane as the other numbers, but perpendicular to number as such. If a number signifies something countable, a zero signifies that there is nothing countable there. You cannot count zero; you can only take account of its uncountability.

Kant is intrigued by another aspect of the joke relation, namely the communication in it of two kinds of thing: representation and the body. Kant tells us that 'this very reduction [of an expectation to nothing] at which certainly understanding cannot rejoice, is still indirectly a source of very lively enjoyment for a moment. Its cause must consequently lie in the influence of the representation upon the body, and the reciprocal effect of this on the mind.'[6] So laughter is not just a violent alternation of contraction and dilation in the muscles, but an alternation between physical muscles and, so to speak, the muscles of the mind involved in forming expectations:

> it is readily intelligible how the sudden act above referred to, of shifting the mind now to one standpoint and now to the other, to enable it to contemplate its object, may involve a corresponding and reciprocal straining and slackening of the elastic parts of our viscera, which communicates itself to the diaphragm (and resembles that felt by ticklish people), in the course of which the lungs expel the air with rapidly succeeding interruptions, resulting in a movement beneficial

> to health. This alone, and not what goes on in the mind, is the proper cause of the gratification in a thought that at bottom represents nothing.[7]

On Kant's account, it is an interchange between the body and the mind, the actual and the represented, form and information, that produces laughter. Kant is followed in this strange economy that connects the physical and the mental by Freud, who, in his *Jokes and their Relation to the Unconscious* (1905), presents his version of the relief theory proposed by Kant and others before him. Freud proposes that, instead of an expectation suddenly being deflated, the mechanism of what he calls the 'joke-work'[8] – in parallel with the 'dream-work' that he had introduced five years previously in *The Interpretation of Dreams* – produces an expectation of an effort of inhibition or repression, which is suddenly removed. Laughter, for Freud, is not a mere incongruity, a friction or tickling of difference, but a sudden alternation of quantity. A quantity of what? we may enquire. For no effort seems in fact to be made here, only an anticipation or feint of an effort. And yet this potential effort nonetheless seems capable of producing a saving or bonus, when it turns out not to be required. Perhaps the equation for this might be written as: $0 + (1 - 1) = 1$. Freud tells a joke, though he is not sure whether it should really count as one, which seems to enact this nonsense economy:

> A gentleman entered a pastry-cook's shop and ordered a cake; but he soon brought it back and asked for a glass of liqueur instead. He drank it and began to leave without having paid. The proprietor detained him. 'You've not paid for the liqueur.' 'But I gave you the cake in exchange for it.' 'You didn't pay for that either.' 'But I hadn't eaten it.'[9]

WORDS AND NUMBERS

Numbers stand out against words. Numbers and words belong to drastically different orders. This is nicely illustrated by the joke about a man who goes to a monastery where all the jokes have been told so many times that they have been assigned numbers. He says a few numbers at random, and is gratified by polite chuckles from all round the room. When however he ventures on the number 367, the room suddenly erupts into laughter, the monks slap each other on the back, clutching themselves and wheezing with helpless laughter. When the guffawing eventually subsides, the man asks his guide why 367 was so much funnier than the jokes indicated by other numbers. 'We hadn't heard that one before,' he replies.

So this suggests a strange antinomy. Words and numbers connote different kinds of value. Words embody values; they are our way of articulating difference of values. No word is equivalent to any other word. Words embody, that is to say, the principle of the incommensurability of values. Numbers, on the other hand, allow for the possibility of equivalence. Any number, as we saw in Chapter Three, can be rendered exactly and entirely in terms of other numbers; indeed, this is the only way in which a number can be defined. Words mean uniqueness: numbers mean equivalence.

Numbers and words appear to have been pulling apart from each other for some time. And yet there is no number that cannot be articulated as a word, or words, nor any mathematical function that cannot in the end be made articulate in words. Contrariwise, we know that every word can now be represented in digital form. So, although words and numbers seem incommensurable, they also in fact interpenetrate; words enclose numbers entirely, and numbers coincide exactly with words. Standing over against numbers, words yet take issue with themselves. I want to show in this chapter that laughter involves, I dare not yet say invariably derives from, this perturbation, from the seemingly alien presence within language of the kind of indifference or equivalence represented by number.

As we saw in Chapter Three, Elizabeth Sewell shows that the playfulness that is characteristic of nonsense writing depends on the two leading characteristics of number, namely distinctness and seriality; numbering assumes and instances a world of absolutely distinct units, and also assumes and instances the arrangement of those units in a series marked by counting.[10] These two principles are so tightly bound together in the simplest mathematical procedures that we do not often notice that they pull in different directions. For seriality embodies absolute incommensurability, since no number can equal another number that occupies a different place in the series; two can never equal three, and four can never equal five. The principle of seriality ensures, not only that all numbers are absolutely distinct, but that all numbers are absolutely unique.

But the principle of seriality also decrees that all numbers are unique, and therefore absolutely distinct from each other, in exactly the same way; that is, they all differ from each other in terms of the units that constitute them. There are three ones in three, and four ones in four; and the 'ones' in each case are absolutely identical and interchangeable. Imagine if, counting from one to four, one had to remember that the intervals between one and two and two and three and three and four were slightly different, and so had to be kept in the right order. But it does not matter a bit what kind of ones are, as we say, 'added up together', since all the 'ones' in question, indeed all 'ones' of any kind, are all the same. So there is no real 'up', since one can add numbers in any direction. In fact, the capacity to order numbers serially, the capacity to count, and therefore the quality which numbers seem to have of allowing or mandating a world of mere numbers, is borrowed from the ordering operation performed upon numbers by the naming of numbers as numerals, or number-words.

There is a story told of the young Carl Friedrich Gauss that may dramatize this tension between the serial and the reversible. His class was asked by their teacher to add together all the numbers between one and a hundred. His peers set about this task with pencil and paper, no doubt most of them ordering the numbers in addition columns: 1 + 2 + 3 + 4, and so on. Carl reflected for a moment or two, then put up his hand. '5,050,' he said.[11] Where the other children had set out to work through the numbers, Carl, possessed of a highly developed capacity to envisage numbers as physical things, had simply 'looked at' the line of numbers from 1 to 100, and recognized that the best place to begin was not at the beginning, but in the middle. Or rather, just after the middle, for he saw that 50 sat next to 51, which, added together, made 101. And, if one took the two numbers that bracketed this pair, 49 and 52, they too added up to 101. And so did the next two numbers out, 48 and 53, just as every other pair would have to, all the way to 2 and 99 and finally 1 and 100. And, since there were exactly fifty such pairs, the required total must be $50 \times 101 = 5050$.

Gauss had performed a calculation by resisting seriality, that is, recognizing the indifference to order of the units ordered in the number line. The number line from 1 to 100 will have many numbers that will seem to a non-mathematical intelligence and indeed to some kinds of mathematicians to be full of erogenous zones and hotspots, numbers possessed of particular kinds of significance, no doubt in part because this particular sequence seems to mark the practical limits of the number of years a human being is likely to live. The numbers between 1 and 100 seem possessed of a certain life, a quality that is unevenly distributed across them, the quality of *being* unevenly distributed, because they serve so well to count up the years of a life. There might be other reasons for according magical associations to numbers: one might equally live at number 76, or regard thirteen as unlucky. All numbers are equal, but, viewed as most human beings do view them, some are more equal than others: because we operate a decimal system, no doubt founded on the convenience of counting on our fingers, tens seem to provide breakpoints or caesuras, octaves (if I may mix my numerical bases for a moment), in the scale.

This tension between distinction and indistinctness is embodied in the distinction between numerology and numerality. Gauss could see past this clumped or lumpy quality of the number line, which can ordinarily only be smoothed into commensurability by the work of calculation, painfully decomposing 17 or 73 to their constituent units. He not only knew, he could, so to speak, see immediately that numbers were all absolutely the same. He could see past the differences in quality of numbers to their indifferent equality. This meant that he was able to break the mesmerizing spell of the number line itself. It did not make any difference where one started, except that there was one point in the sequence, a sort of cardinal point, just around the middle, where this principle was best illustrated, so that calculation was scarcely needed at all.

Samuel Beckett has the character Arsene in his most mathematical novel *Watt* voice something of this same equanimity. Arsene is about to leave the house of Mr Knott, and is delivering himself of a peroration in which he attempts to provide some account of his time in the house and what he has learned from it. And what he has learned is precisely that there is nothing, or nothing cumulative, to be learned:

> And if I could begin it all over again, knowing what I know now, the result would be the same. And if I could begin again a third time, knowing what I would know then, the result would be the same. And if I could begin it all over again a hundred times, knowing each time a little more than the time before, the result would always be the same, and the hundredth life as the first, and the hundred lives as one. A cat's flux. But at this rate we shall be here all night.[12]

I am not sure that this is exactly a joke in itself, but there is something joke-like in its structure, consisting as it does of an open accumulation of verbal circumstance that rounds up, or down, to nothingness. But perhaps there is some significance in that 'exactly a joke'; perhaps everything I will be saying in this chapter may amount or be reduced to the observation that when a joke is almost a joke, but not quite, it is not really a joke at all, and when it is, it is absolutely. Comedy is all-or-nothing digital; tragedy is that'll-do analogue.

Dickens is often represented as a writer of imaginative excess, a writer who, in the prodigiousness of his invention, spills exuberantly beyond measure and proportion. Dickens set his face against the grim hedonic calculus of what he took to be utilitarianism (among the few things for which Dickens daily needs absolution is the vicious and stupid misunderstanding of utilitarian philosophy he bequeathed to a literary culture that remains smug and ignorant about it), promoting the principles of disproportion and the measureless. Dickens's critique of utilitarianism is embodied in the figure of Thomas Gradgrind in *Hard Times*, who is introduced to us as

> A man of realities. A man of facts and calculations. A man who proceeds upon the principle that two and two are four, and nothing over, and who is not to be talked into allowing for anything over . . . With a rule and a pair of scales, and the

> multiplication table always in his pocket, sir, ready to weigh and measure any parcel of human nature, and tell you exactly what it comes to. It is a mere question of figures, a case of simple arithmetic.[13]

And yet Dickens himself was a writer who, in his successful exploitation of serial fiction, lived and wrote, literally, by numbers, in thrall to the endlessly renewed demand that he fill up the 32 printed pages that were required for each monthly part of the novels he wrote over twenty months. Dickens thrived on excess, but it was an excess that he subjected to mathematical control, and that was tightly dependent on mathematical constraints for its quantitative easings. He liked to measure the success of his legendary public readings by the number of ladies who were carried out insensible. And Dickens's comedy, like his writing practice in general, is in fact intertwined and impregnated with number from top to bottom.

As many have observed, Dickens's comedy often depends upon the reduction of character to a single trait or mechanical mannerism. In the case of Uncle Pumblechook, who is one of the many tormentors of the infant Pip in *Great Expectations*, it is his compulsion to keep Pip up to the mark by means of continuous arithmetic.

> I considered Mr Pumblechook wretched company. Besides being possessed by my sister's idea that a mortifying and penitential character ought to be imparted to my diet – besides giving me as much crumb as possible in combination with as little butter, and putting such a quantity of warm water into my milk that it would have been more candid to have left the milk out altogether – his conversation consisted of noth-ing but arithmetic. On my politely bidding him Good morning, he said, pompously, 'Seven times nine, boy?' And how should I be able to answer, dodged in that way, in a strange place, on an empty stomach! I was hungry, but before I had swallowed a morsel, he began a running sum that lasted all through the breakfast. 'Seven?' 'And four?' 'And eight?' 'And six?' 'And two?' 'And ten?' And so on. And after each figure was disposed of, it was as much as I could do to get a bite or

> a sup, before the next came; while he sat at his ease guessing nothing, and eating bacon and hot roll, in (if I may be allowed the expression) a gorging and gormandizing manner.[14]

Two orders are brought into collision here. First of all, there is the order of eating, measured, as so often in Dickens, with an alternating economy of generosity and niggardliness. Where the kindly, helpless blacksmith Joe spoons gravy on to Pip's plate in recompense for the domestic humiliations he must suffer, Pumblechook's homeopathic dilution of the milk of human kindness makes for subtraction where increase should be. Then there is the alternative order of calculation, which, through Pumblechook's renewed inquisition, monopolizes the organ of eating, the mouth, which is thereby reduced to a round, empty zero. The ongoing calculation scarcely deserves the name of mental arithmetic, since its effect is to replace eating with inanition, rations with rationality.

It would be easy to cash out the comedy of this passage through a Bergsonian analysis that would see it as taking revenge on Pumblechook by lopping him down to his impulse to impose single-minded and sadistic arithmetic. The more Pumblechook piles on the numbers, the more he is himself reduced to a single characteristic, to a unity of being that is in fact only an unnatural fraction of what it ought to mean to be a human being. The notable fact here though is that, as always in such cases, Dickens enters so far into Pumblechook's maniacal mathematics in order to achieve his comic effect. Dickens's narrative plays with the possibility that it might itself get caught up in the churning gears of Pumblechook's arbitrary arithmetic. Pumblechook's improvised tot has no answer or outcome, and the numbers both do and do not matter. If Pumblechook reduces Pip to a number-crunching machine, he and his narrative are caught in the jaws of the same logic. Indeed, Pumblechook is represented increasingly as subjected to the process with which he seeks to subjugate Pip:

> we came to Miss Havisham's house, which was of old brick, and dismal, and had a great many iron bars to it. Some of the windows had been walled up; of those that remained, all

> the lower were rustily barred. There was a courtyard in front, and that was barred; so we had to wait, after ringing the bell, until some one should come to open it. While we waited at the gate, I peeped in (even then Mr. Pumblechook said, 'And fourteen?' but I pretended not to hear him).[15]

Dismissed from the gate by the pert Estella, Pumblechook attempts to regain some of his crumpled dignity with a parting bit of moralism:

> [he] departed with the words reproachfully delivered: 'Boy! Let your behavior here be a credit unto them which brought you up by hand!' I was not free from apprehension that he would come back to propound through the gate, 'And sixteen?' But he didn't.[16]

Nevertheless, the running joke of Pumblechook's running sum is brought to a kind of reckoning, when Pip returns from Miss Havisham's and is reluctant to reveal what has occurred there:

> 'First (to get our thoughts in order): Forty-three pence?'
>
> I calculated the consequences of replying 'Four Hundred Pound,' and finding them against me, went as near the answer as I could – which was somewhere about eightpence off. Mr Pumblechook then put me through my pence-table from 'twelve pence make one shilling,' up to 'forty pence make three and fourpence,' and then triumphantly demanded, as if he had done for me, '*Now*! How much is forty-three pence?' To which I replied, after a long interval of reflection, 'I don't know.' And I was so aggravated that I almost doubt if I did know.
>
> Mr. Pumblechook worked his head like a screw to screw it out of me, and said, 'Is forty-three pence seven and sixpence three fardens, for instance?'
>
> 'Yes!' said I. And although my sister instantly boxed my ears, it was highly gratifying to me to see that the answer spoilt his joke, and brought him to a dead stop.[17]

Pumblechook thinks to have finished Pip off with his inquisition, but it is really Pip who, in the detail slyly insinuated by his author, has 'calculated the consequences'. By refusing to take the sum seriously, Pip exposes Pumblechook to the indifference of number that he has himself been wielding as a weapon. The principle that Pumblechook brings to bear on Pip, namely of reducing everything to number, is itself applied to him. There are two competing orders of arithmetic, just as there are two jokes: Pumblechook's and Pip's, which 'spoilt his joke, and brought him to a dead stop'.

At the beginning of his book on laughter, Bergson helps us to recognize an important aspect of this kind of satirical humour, when he points to the strange neutrality that is essential to the comic impulse:

> I would point out . . . the absence of feeling which usually accompanies laughter. It seems as though the comic could not produce its disturbing effect unless it fell, so to say, on the surface of a soul that is thoroughly calm and unruffled. Indifference is its natural environment, for laughter has no greater foe than emotion . . . the comic demands something like a momentary anesthesia of the heart. Its appeal is to intelligence, pure and simple.[18]

We may perhaps describe feelings as the embodiment of values: feelings are the way in which we enact the fact and the manner of things mattering to us. The equatability or equivalence of all values that is characteristic of the numerical suggests to Bergson a world without feeling, a world of pure intelligence:

> In a society composed of pure intelligences there would probably be no more tears, though perhaps there would still be laughter; whereas highly emotional souls, in tune and unison with life, in whom every event would be sentimentally prolonged and re-echoed, would neither know nor understand laughter.[19]

The work in which Beckett comes closest to immersing himself and his reader in the destructive equanimity of number is surely *Watt* and, within that novel of obsessive accumulations, permutations and calculations, the most sustained exercise in mathematized narrative is the episode, allegedly recounted by Arthur to Watt and others in Mr Knott's garden, which deals with the appearance before a College committee of Ernest Louit, accompanied by what he claims to be a mathematical savant from the far West of Ireland, in order to account for the £50 of college funds that he has expended in research for the dissertation he entitles 'The Mathematical Intuitions of the Visicelts'. Louit has plainly put the research grant advanced to him by the College to other uses than the investigation of mathematical capacities among the indigent indigenes of County Clare (for the amazement of whom £5 has been set aside in his budget to purchase 'coloured beads'). In a sense, the entire episode is an attempt to supply an alternative, and extravagantly inflationary, budget in place of the simple account of how the money has in fact been spent. Numbers begin early on to take the place of words:

> The College Bursar now wondered, on behalf of the committee, if it would be convenient to Mr. Louit to give some account of the impetus imparted to his studies by his short stay in the country. Louit replied that he would have done so with great pleasure if he had not had the misfortune to mislay, on the very morning of his departure from the west, between the hours of eleven and midday, in the gentlemen's cloakroom of Ennis railway-station, the one hundred and five loose sheets closely covered on both sides with shorthand notes embracing the entire period in question. This represented, he added, an average of no less than five pages, or ten sides, per day. He was now exerting himself to the utmost, and indeed he feared greatly beyond his strength, with a view to recuperating his MS., which, qua MS., could not be of the smallest value to any person other than himself and, eventually, humanity.[20]

Numbers begin also to infiltrate the account of the enquiry, first of all in the account provoked by the seemingly innocuous statement that 'The committee . . . began to look at one another,' followed immediately by the odd and ominous qualification, 'and much time passed, before they succeeded in doing so'. As so often in *Watt*, a simple proposition detonates a long chain of permutational reasoning:

> when five men look at one another, though in theory only twenty looks are necessary, every man looking four times, yet in practice this number is seldom sufficient, on account of the multitude of looks that go astray. For example, Mr. Fitzwein looks at Mr. Magershon, on his right. But Mr. Magershon is not looking at Mr. Fitzwein, on his left, but at Mr. O'Meldon, on his right. But Mr. O'Meldon is not looking at Mr. Magershon, on his left, but, craning forward, at Mr. MacStern, on his left but three at the far end of the table. But Mr. MacStern is not craning forward looking at Mr. O'Meldon, on his right but three at the far end of the table, but is sitting bolt upright looking at Mr. de Baker, on his right. But Mr. de Baker is not looking at Mr. MacStern, on his left, but at Mr. Fitzwein, on his right. Then Mr. Fitzwein, tired of looking at the back of Mr. Magershon's head, cranes forward and looks at Mr. O'Meldon, on his right but one at the end of the table.[21]

The only solution to the irrational waste and blunder of all these wildly misdirected eyebeams, our insanely meticulous author tells us, is for the committee to mathematize the process of looking at itself, by assigning each committee member a number:

> Then, when the time comes for the committee to look at itself, let all the members but number one look together at number one, and let number one look at them all in turn, and then close, if he cares to, his eyes, for he has done his duty. Then of all those members but number one who have looked together at number one, and by number one been looked at one by one, let all but number two look at number two, and let number two in his turn look at them all in turn, and then remove, if his eyes are sore, his glasses, if he is in the habit of wearing glasses, and rest his eyes, for they are

> no longer required, for the moment. Then of all those members but number two, and of course number one, who have looked together at number two, and by number two been looked at one by one, let all with the exception of number three look together at number three, and let number three in his turn look at them all in turn, and then get up and go to the window and look out, if he feels like a little exercise and change of scene, for he is no longer needed, for the time being. Then of all those members of the committee with the exception of number three, and of course of numbers two and one, who have looked together at number three and by number three been looked at one by one, let all save number four look at number four, and let number four in his turn look at them one after another, and then gently massage his eyeballs, if he feels the need to do so, for their immediate role is terminated. And so on, until only two members of the committee remain, whom then let at each other look, and then bathe their eyes, if they have their eyebaths with them, with a little laudanum, or weak boracic solution, or warm weak tea, for they have well deserved it. Then it will be found that the committee has looked at itself in the shortest possible time, and with the minimum number of looks, that is to say x squared minus x looks if there are x members of the committee, and y squared minus y if there are y.[22]

The text depends upon the tussle between words and numbers. This might approximate to a contest between temporality, for words, at least in their condition as utterance, must transpire in time, and spatiality. Actually, we should acknowledge that this is a conflict which exists within mathematics, in the relation between the formula and the proof, the *quod* and the *demonstrandum*. Unless, perhaps, mathematics is nothing else but this tension between what is and the working out of what is; God, outside time, presumably does not do mathematics, since he knows the answers already – though Wittgenstein wonders 'Can God know all the places of the expansion of π?'[23] The work of *Watt* is to rotate the mathematically simultaneous into the wordy dimension of the consecutive, and then back again. Linearity, by which is really meant irreversibility, is repeatedly folded back into reversibility. The filling of space by oscillation and alternation takes the place of onward movement from one point to another, meaning that the space of the novel is filled rather than traversed. Numbers are able to order the world in the way they do precisely because they make the order, in the sense of the order of succession of things, irrelevant. This process is announced in *Alice's Adventures in Wonderland*:

> 'What do you know about this business?' the King said to Alice.
>
> 'Nothing,' said Alice.
>
> 'Nothing *whatever?*' persisted the King.
>
> 'Nothing whatever,' said Alice.
>
> 'That's very important,' the King said, turning to the jury. They were just beginning to write this down on their slates, when the White Rabbit interrupted: 'Unimportant, your Majesty means, of course,' he said, in a very respectful tone, but frowning and making faces at him as he spoke.
>
> 'Unimportant, of course, I meant,' the King hastily said, and went on to himself in an undertone, 'important – unimportant – unimportant – important – ' as if he were trying which word sounded best.[24]

The Louit episode in *Watt* is full of doubled and multiplied words: 'yes yes', 'no no', 'haha', 'come come', 'oh no no no no no'. Words are not numbered, but numerous. And the principle of reversibility is also powerfully in evidence. Beckett's drafts indicate that Mr Nackybal's name is a derivation from Caliban, itself of course an adjustment of Cannibal. Nackybal is converted in the episode to Ballynack and Nackynack. Cannibalism seems to be a metaphor for the churning of elements in the episode. Louit explains that hunger has forced him to roast and eat his faithful dog O'Connor, leaving only his bones, and Beckett probably only refrains with difficulty from concluding this story with the traditional Irish-bull ending, which would have had Louit lamenting: 'A pity O'Connor isn't here; he'd have loved these bones.'[25]

Our humanistic prejudices may incline us to say that words are here being reduced to the inhuman definiteness of numbers, but Beckett's text is determined to show us something like the reverse, that to numerize is to defer the possibility of making any final or finite statement. Insofar as it passes through number, the pursuit of completeness or absolute truth will always be put at infinite risk.

The arts, self-identifying as they, possibly we, are with the indefinite, the open and the fluidly non-absolute, are inclined to view scientific reasoning as paralysed by abstraction and a kind of false, inhuman positivity. In fact, though, there are reasons to suspect the arts of confusing absoluteness with exactitude. It is in fact approximation that allows for absoluteness. This is well illustrated in the story of the researcher seeking responses from different kinds of academic to the suggestion that all odd numbers are primes. The mathematician says: '1 is prime, 3 is prime, 5 is prime, 7 is prime, 9 is not prime. The conjecture is false.' The physicist says: '1 is prime, 3 is prime, 5 is prime, 7 is prime, 9 is not prime, 11 is prime, 13 is prime. Within acceptable limits of measurement error, the conjecture holds.' The literary critic says: '1 is prime, 3 is prime, 5 is prime, 7 is prime, 9 is prime – it's true! All odd numbers *are* prime!'

Mathematics has the reputation of being more economical and less wasteful than words, but it is words that encourage us impatiently to square things off and round things up into always approximative absoluteness. Words save time, the time that numbers are. But at this rate we shall be here all night.

LAUGHING BY NUMBERS

My proposal is that laughter is produced from the friction and fission of the positive values that are put into play by the joke-work and the pure negativity, negativity that is best embodied by number, that intersects with them. Things that matter suddenly come to nothing, are suddenly made to be things that do not matter at all; nonequivalence is rotated suddenly into absolute equivalence. Equivalence is not just nothingness. It can also be considered as a kind of null infinity, for the equivalence of numbers, their capacity to be manipulated and reversed and recombined, means that there is no end to their equivalences, which are therefore indifferently everything and nothing.

The most surprising reversal in this is that it is now the order of words that signifies the positivity of meaning, or value. The order of numbers, by contrast, signifies, not the quantifiable, but the nonquantifiable, the nothing-at-all that is equivalent to anything-at-all. So, unexpectedly, it is number that represents the eruption of the nonquantical into the order of the quantical, of equality into quality. Just as the sign for zero seems to be the intersection of the order of numbers and the nonnumerical, so numbers can act as the intersection of the positive qualities signified by words, and the indifference of number. If nothing is the other of number, then number is the nothingness that is the other of words, that nevertheless is powerfully at work within words. Number is the other of words that words themselves harbour, with hilarity the outcome of its demonstration.

G. H. Hardy himself suggests something of this near-nihilism that exists within number, in his final estimations of the value of his own mathematical life:

> I have never done anything 'useful'. No discovery of mine has made, or is likely to make, directly or indirectly, for good or ill, the least difference to the amenity of the world. I have helped to train other mathematicians, but mathematicians of the same kind as myself, and their work has been, so far at any rate, as I have helped them to it, as useless as my own. Judged by all practical standards, the value of my mathematical life is nil; and outside mathematics it is trivial anyhow.[26]

Others might be inclined to see this as a claim for the intrinsic value of an intellectual enterprise, rather than its instrumental value, but it is notable that Hardy insists that this almost-nullity is nevertheless to be measured numerically:

> The case for my life, then, or for that of anyone else who has been a mathematician in the same sense in which I have been one is this; that I have added something to knowledge,

> and helped others to add more; and that these somethings have a value which differs in degree only and not in kind, from that of the creations of the great mathematicians, or of any of the great artists, great or small, who have left some kind of memorial behind them.[27]

Our complacent assumption is that laughter has something to do with our triumph over the inert, that it is life asserting its claims against the givenness or dead necessity of things. But the strong implication of the mathematically driven comedy I have considered here suggests that this cannot be the whole story. For the implication of number and pure quantity in comedy suggests that it must at least include some insurgence of the inert, an assertion of the purely quantical against the world of quality. We do not merely laugh at number, we also laugh by numbers. To grasp this properly, we need to recognize that number is itself plural. There is the kind of number we use to count with, and therefore to assign values, for example by maintaining the difference between the one and the many. This is number in the service of difference, number that we can count on. But we have seen that there is another dimension of number. This is the giddiness of number as pure, unrelieved and, so to speak, indifferent differentiation. To be sure, there is a kind of dissolute exhilaration in this indifference, but we saw in Chapter Three that there is a horror too – the horror of losing count, of being given over to number without being able to count on it. This is not life asserted against death, but death come uncountably, unaccountably, to life. It is not death driven back by life, but life inundated by the death of the indifferent. Laughter is not gaiety in the face of death, but death itself made facetious.

Beckett's laughter appears like a relief from what surrounds it, and therefore the guarantee that things are not really as serious as all that. But the refusal to contain laughter comes to the same thing as the refusal to allow it. This is not a laughter that punctures logic, but one that steps into its place. A text like *Watt*, which becomes a mechanical laughter-machine, an algorithmic *risus sardonicus* that does not undo death but rather does its grim, grinning work, is only the most extreme demonstration of a general characteristic of comic writing.

We work so hard at laughter in order to overcome it, rather than to overcome with it. If laughter comes from the eruption of nothing in the place of something, laughter is also the defence against the propagation of this nothing. Where laughter propagates, we seek through our routines of comedy – in the regulated rhythms of the joke, for example – to contain, drain and exhaust it. We laugh to fend off death by laughing, we laugh to have done with laughing, in a controlled explosion rather than a general conflagration. Laughter is a locking together of the machinery that laughing itself looses. That is why laughter, apparently and allegedly the dissolution of power, in fact usually works to solidify and concentrate it. Laughter is less colonic irrigation than colonial occupation. A wise lecturer takes care to laugh his lecturees into concupiscent acquiescence. The urge to pass a joke on is the urge of the crowd to become more of a crowd, to exclude nothing that it does not already contain.

MEASURES OF PLEASURE

Though laughter characteristically gives pleasure, and is usually associated with it, laughter and pleasure are not entirely identical. Indeed, there is a calculative aspect to this relation: beyond a certain point, being helpless with laughter can come close to pain. Pleasure has the reputation for being spontaneous, unreflective. But, if the mechanics of jokes make the quantical aspects of laughter-production particularly evident, this should not blind us to the fact that pleasure must also be *taken*, which is to say, taken account of. Dickens captures this very well in *Martin Chuzzlewit*, in a scene in which the character Tom Pinch sits in a bar in Salisbury:

> All the farmers being by this time jogging homewards, there was nobody in the sanded parlour of the tavern where he had left the horse; so he had his little table drawn out close before the fire, and fell to work upon a well-cooked steak and smoking hot potatoes, with a strong appreciation of their excellence, and a very keen sense of enjoyment. Beside him, too, there stood a jug of most stupendous Wiltshire beer; and the effect of the whole was so transcendent, that

> he was obliged every now and then to lay down his knife and fork, rub his hands, and think about it.[28]

That this is a comic moment might aptly remind us of Freud's complex negotiations with the economies of laughter. We have seen that the essence of the Freudian theory of the comic, which significantly is not focused on the unbound energies of spontaneous hilarity, but on the geared machinations of the joke, is that it involves the differential investment of quantities of psychic labour, in order to manufacture a tension that can be profitably released as laughter. This is a local application of the general economic principle in Freud that there is no pleasure without the overcoming or outflanking of obstacles.

Pleasure is never simple – it is always in fact duplicitous (at least). In this it does not resemble pain, which is immediate, self-announcing and self-interpreting. Pain can be diffuse, and difficult to describe, but there is rarely any doubt that it is there. Perhaps this is why pain has so often been thought of as the guarantee of the real, for example in Fredric Jameson's stern announcement that 'History is what hurts. It is what refuses desire and sets inexorable limits to individual as well as collective praxis.'[29] Pleasure and the desire for it seem, by contrast, much less certain or self-evident, and much more intrinsically difficult simply to experience. You seem to have, like Tom Pinch, to give it some thought.

Whereas pain saturates and yet also abolishes time, pleasure is intimately intermingled with temporal experience, from which it borrows. The anticipation of pleasure, even if it is in the negative form of the retreat of pain, gives tone, texture, tension and extenuation to time. It is hard to think of a pleasure that does not involve or require some kind of protention or retrospection. So, where pain is absolute, pleasure is relative. Pleasure can never be wholly *en-soi*, or in itself, it must always be in part *pour-soi*, for itself, which is to say, it must involve some minimal form of reflexivity. It is for this reason that it seems to make sense to say that one might suddenly become aware that something is or was pleasurable, but it does not seem easy to conceive how one might be in pain without realizing it. One might say that pleasure is never fully aware of itself, that it is not until it is represented in some way that it can be recognized as pleasure. Pleasure is always compound in form, that is, it always involves some kind of estimation or taking stock. This implies that travail or difficulty are not the opposites of pleasure, but part of its repertoire.

One of the principal forms of pleasure's reflexivity is its self-subjection to metric and quantification. The conjoining of work, pleasure and number is nowhere better evidenced than in the development of modern sport. Before the massive and ramified codification of sports that took place in the later nineteenth century, almost exclusively in England, not insignificantly the most advanced industrial nation in the world at that point, little account was taken of measurement or scoring in sport.[30] A medieval football match between two villages, which might last all day and lead to many broken heads and limbs, ended when one team scored. There was no opportunity to go for an equalizer, no change of ends, no best-of-three or penalty shoot-out. Victory was absolute, crushing and final. Though we may try to pretend that we have lost something vivid and precious in the replacement of the all-or-nothing excess of carnival sports by rules and scoring systems, in fact this is part of a process of redistributing the capital of pleasure. The pleasures of the excessive, immoderate and transgressive, as promoted influentially, for example, in the work of Georges Bataille, are often, perhaps usually, both anti-democratic and aristocratic.

A contrast is detectable between pleasure and pain in respect of measure. Where measurement can help to control or diminish pain (assigning to a particular pain a value on a scale running from one to ten, for example), measurement can be and is commonly used to intensify and prolong pleasure. The work of Freud provides some of the richest material on the economic structures that pleasure employs, or that subjects employ to get or keep or manage pleasure. Measurement is a form of management.

That the economics of pleasure can extend far beyond the co-ordination of work and leisure is demonstrated in Peter Sloterdijk's arguments in his *Rage and Time* (2010) for the constitutive role of anger management in twentieth-century politics. Anger, he points out, tends to ecstatic but politically wasteful eruptions of tension-reducing violence. If anger – the resentment of the proletariat at their systematic immiseration, say – is to be turned to political account, it must be concentrated and coordinated. Anger is a kind of affective capital that must be accumulated in anger-banks. This borrows from the logic of Christian eschatology, which requires the put-upon soul to suffer and be still, keeping their anger in trust with the Deity, who, declaring 'Vengeance is mine,' monopolizes all anger until the final apocalyptic purgation of the Day of Wrath. But now it is the Party or the State that monopolizes the right to express anger, but must also ration it, in order to keep the anger-banks well stocked. The pleasure and unpleasure of anger are subject to the most complex kinds of coordination.[31]

We have convinced ourselves, especially the we that is presupposed and presumed upon in a gathering of non-mathematical persons, that, whatever else it may do, the mathematization of the world must pose a threat to our humanity and freedom. 'I am not a number,' roars the Patrick McGoohan character at the beginning of each episode of the 1960s cult series *The Prisoner*, 'I am a free man.' If there is one solidary article of faith among those in the humanities, it is that there is a deep and dangerous antagonism between the realm of number and the realm of words and images. The realm of the qualitative must be secured against the deadening incursions of the quantitative. The presence or prominence of number is the great discriminator between the sciences and the humanities. The more the realm of number expands, we fear, and thereby also reassure ourselves, the more the realm of the human diminishes. We know it, we are sure of it, we have no need to think about it any more, indeed we cannot waste or risk time thinking about it, lest we cease to be able to think with it. But might one risk the suggestion that we find it hard to associate pleasure and number because we are so many of us casualties of an educational system, sustained by a long and blundering set of prejudices, that failed to make this association possible?

In his *Art of Discovery* of 1685, Leibniz looked forward to the day when calculation might take the place of disputation: 'The only way to rectify our reasonings is to make them as tangible as those of the Mathematicians, so that we can find our error at a glance, and when there are disputes among persons, we can simply say: Let us calculate [*calculemus*], without further ado, to see who is right.'[32] A defining strain preparing for the contemporary allergy to number in the humanities is the Romantic protest against the powerful efforts to put social and political reasoning on a firm basis by employing calculative reason, especially in the philosophical form of utilitarianism, so influentially attacked by Dickens.

Jeremy Bentham was not the first, but was certainly the most systematic and influential exponent of utilitarian philosophy, that is, the philosophy that insists that the value of anything is to be defined wholly and without residue in terms of its utility, or its tendency to produce 'benefit, advantage, pleasure, good, or happiness'.[33] As an unrepentant utilitarian, if also, I hope, a more versatile one than Thomas Gradgrind, I must declare myself one of Bentham's sect. Quantity and measurement are at the heart of utility, since the utility of an action or idea arises when 'the tendency it has to augment the happiness of the community is greater than any which it has to diminish it' (*Introduction*, 3). Bentham did in all earnestness, and to the quick derision of many, propose what he called a 'felicific calculus' that would allow one to calculate the exact quantity of pleasures and pains. Since Bentham's single governing moral principle was the production of pleasure and the reduction of pain for the greatest number, such an effort at quantification was unavoidable.

The felicific calculus is set out in Chapter Four of his *Introduction to the Principles of Morals and Legislation* (1789), which is entitled 'Value of a Lot of Pleasure or Pain, How to Be Measured'. There he distinguished seven different dimensions of the pleasures or unpleasures that might be produced by a given action. These dimensions were: 1) the intensity of the pleasure or pain; 2) its likely duration; 3) its certainty or uncertainty; 4) its propinquity or remoteness; 5) its fecundity, by which Bentham means 'the chance it has of being followed by sensations of the same kind'; 6) its purity, that is 'or the chance it has of not being followed by sensations of the opposite kind'; and, finally, 7), its extent, that is, the number of persons whom it may affect (*Introduction*, 30). Bentham even produced a mnemonic jingle (not *very* mnemonic) to help his students keep this algorithm in mind:

Intense, long, certain, speedy, fruitful, pure –
Such marks in *pleasures* and in *pains* endure.
Such pleasures seek if *private* be thy end:
If it be *public*, wide let them *extend*.
Such *pains* avoid, whichever be thy view:
If pains *must* come, let them *extend* to few. (*Introduction*, 29)

Moral reflection is thereby reduced – or maybe raised – to mathematical reasoning. The usual, and right, thing to say about the hedonic calculus is that it is impossible to do the sums. The usual, but erroneous, thing to say about why this is so is that pleasure and number are inimical, or that pleasure is unquantifiable. It is true that one of the real problems with Bentham's calculus is its presupposition of some common measure or single currency, which would allow one in principle to add and subtract between these different qualities. Bentham was frank in his acknowledgement that the closest approximation we have to this common measure is money. It is perhaps for this reason that utilitarian philosophers have sometimes adopted the terms hedons and (inspired coinage) dolors for the units of felicific currency. In the course of a discussion of the proportioning of offences and punishments, Bentham deals with the objection that 'passion does not calculate,' to which his sturdy and straightforward response is that it is not true. The Benthamite reply to the objection that we cannot quantify pleasure is simply that we so manifestly and continuously do:

> When matters of such importance as pain and pleasure are at stake, and these in the highest degree (the only matters, in short, that can be of importance) who is there that does not calculate? Men calculate, some with less exactness, indeed, some with more: but all men calculate. I would not say, that even a madman does not calculate. Passion calculates, more or less, in every man: in different men, according to the warmth or coolness of their dispositions: according to the firmness or irritability of their minds: according to the nature of the motives by which they are acted upon. Happily, of all passions, that is the most given to calculation, from

> the excesses of which, by reason of its strength, constancy, and universality, society has most to apprehend: I mean that which corresponds to the motive of pecuniary interest. (*Introduction*, 187–8)

The real problem with the felicific calculus is not that it enforces calculation where none is possible, but that there are so many ways of doing the calculations. The problem is not that the felicific calculus is too rigid and inapplicable to the circumstances of pleasure, but, as Wesley C. Mitchell observed many years ago, that it is too obliging to them.[34] But this does not diminish the fact that pleasure and measure are in fact tightly intertwined. Far from being the adversary of number, pleasure is, in some ways, its apotheosis.

WE CAN WORK IT OUT

One of the great sources of pain attaching to the question of pleasure is that we persist in thinking of pleasure as the inverse of work. The harder we work, we reason, the less pleasure we have; the less work we do, the more pleasure we will have. We should be watchful whenever we find ourselves reasoning on the basis that anything is the opposite of anything else, but particularly when it comes to pleasure. Since pleasure is the motive principle of everything we do, it finds ways of inveigling itself into everything that seems inimical to it. When I was a student, I had a number of menial jobs in factories and the like, and, though it was not particularly exacting, I found the tedium of the work hugely depressing and fatiguing. Like many another worker, the way I found to reduce these pains was not to shirk and skive, but actually to throw myself into the work. I had, say, to spend all day on the de-burring machine, a lathe-like wheel that removed the rough snags on the side of the little rectangles of copper – destined ultimately to become printed circuits – that another machine had stamped out. I stood at the wheel with a pile of copper rectangles beside me which was almost my height, and my job was to de-burr them. It was easily possible to do four or five of these a minute, but it was very hard indeed to carry on doing four or five a minute for sixty minutes an hour and for seven hours a day. Still, if I set myself the task of doing, say, six or seven a minute in spurts, things changed. Simply varying the number of copper rectangles I managed to de-burr a minute somehow sweetened, yea, de-burred the task itself, just a little, but, given its soul-corroding monotony, a little was more than enough. And, of course, once I realized that I was getting better at the task, and was regularly achieving rates of six or seven a minute, I began to wonder whether I might not be able to do eight for, say, three or even four consecutive minutes. This required vigilance and planning. I needed to bring my performance under Taylorian scrutiny, assessing the ways in which I picked up the copper oblongs, and even the order in which I did them (long side first, or short side first?). I laid wagers with myself, devised inducements and rewards for prolonged good performance. For example, after ten straight minutes of eight a minute, I would have made a profit in copper rectangles of between fifteen and twenty, and therefore in the currency of time of between two and three minutes, which gave me the opportunity to roll and partially consume a cigarette (this was a long time ago, when smoking was a necessary and expected part of any kind of organized labour).

By subjecting my performance to mechanical survey, and calculating outcomes and margins, I escaped the condition of mechanization to which I was otherwise painfully delivered. At the same time I discovered a further source of pleasure in the reckoning itself. Even if I flagged and failed to reach my targets, even if the exhilaration of performing eight breakneck de-burrings a minute began to pall, I had at least the fact of the calculative perspective open to me. I had a critical relation to the work I was doing, a relation that, insofar as it was calculative, was in fact playful. Because I was not reduced to my work, I was in fact no longer alienated from it. It was my work, no longer the work I *had* to do (was required to do), but the work I had to do (was on hand for me to do). I had thus defeated the purpose of the task, as it seemed to me, which was to obliterate any possibility of my being or doing other than the task itself, and to isolate me in the ongoing, outgoing, agonizingly homogeneous, oleaginously oozing present of the work. The conjoining of my exertion with estimation had brought time under tension; it had turned my labour into a project. I was not just calculating my pleasure; I was pleasing myself with my calculations.

All students and newcomers to circumstances in which not very demanding work must be done at a steady rate over a sustained period will sooner or later discover the pleasure of subjecting things to measure in order to vary the beat. And most of those tyros will also sooner or later be made forcibly aware that there is in fact a further, complicating calculation to be made. For I was going to be working in that factory for, at most, six weeks. The people I was working alongside had been in the job for years and much depended for them on their being able to remain in it for many further years (this was a long time ago, when having the same job for years was regarded as a kind of curse). Most of them had found the optimum level of performance, that balanced out all the countervailing pressures and could be sustained, day in and day out, over long periods, even, if necessary, lifelong. Though it might well be possible for them to match the blistering rate of production to which I aspired and which I was able intermittently to attain, I was like a quarter-miler setting the pace for a marathon, and it was not going to be possible for them to maintain that rate for the rest of their working lives. I had to be stopped, and, of course, I was. I was forced, by means of various machinations and ethical humiliations, to ease off, and the work became again, as it had been at the beginning, slow-dripping torture. I had discovered that it was easier to work hard than to take it easy, indeed, that, under many circumstances, ease is agony.

I realized that what mattered was not the quality of life that was achieved and how the chances of it were distributed under a given social arrangement. Nor did it fundamentally matter that the people who did most of the work did not get most of the profit, important though that is. What mattered, for the Romantic Morrisian I became, well before ever reading any of the insipid works of the admirable William Morris, was not the profit that might be made from work, but the quality of the work that it was possible for people to do, or, at least the quality of the relation they had to their work. I was about to go to university to study English, and, whenever I was faced with learning a list of Old English verb inflections, or slogging my way through *The Faerie Queene*, the *Morte D'Arthur*, or, for that matter, the wearisome *News from Nowhere*, I infallibly remembered my hours and days at the de-burring machine, and knew that I was in fact in Paradise, compared with the Purgatory of not having work of such a kind to do. The cruellest social divide, I thought, is not between people who are well and badly rewarded for the work they do, but between people who have work that they would do anyway for nothing, and people who would give almost every penny they earned not to have to do the work they do to earn it.

There are, of course, under some circumstances, pleasures to be had from the remission of or abstention from work. But the receding of work altogether leads to the kind of nightmare that opens up for Philip Larkin in his poem 'Toads Revisited', which recalls and reverses the views of an earlier poem, 'Toads'. The later poem, written in 1962, reflects on why it is that, although 'Walking around in the park/ Should feel better than work', there is a creeping horror in the prospect of 'Being one of the men/ You meet of an afternoon:/ Palsied old step-takers./ Hare-eyed clerks with the jitters,/ Waxed-fleshed out-patients/ Still vague from accidents'.[35] 'Toads' begins 'Why should I let the toad *work*/ Squat on my life?' The sadder, wiser, counterpart poem ends 'Give me your arm, old toad;/ Help me down Cemetery Road.'[36]

Michel Serres evokes the strange interpenetration of hard and soft work in *The Parasite*: 'Work flows from me like honey, like the spider's web . . . I work hard, I don't work at all; it comes easily, just like what an animal does when it follows its own instinct in doing this and that. I am a bee, or a spider, a tree.'[37] I am my work, I am, as the cliché has it, fulfilled by it, because I am not it, which is to say, I have a non-necessary relation to it. My work can fulfil me only when it is not wholly me, or I am not wholly it, so that it can actually release me from the killing condition of having to coincide with myself. Work is nonalienating when it allows me to encounter and enter into my own otherness to myself. When I am not my work, when I am, as we so idly say, alienated from it, when my work is merely what I must do, in order that I have the wherewithal to be able, in some other time and place, after work ceases, to buy back my pawned life, then I am alienated, not from identity, but from this possibility of non-self-coincidence. It is not that one relation to work is qualitative and the other merely quantitative, as we might sluggishly think; it is that one gives the opportunity for complex forms of measure, which is therefore to say, pleasure, and the other does not, beyond the bleak equality of 1 = 1, I am that I am, I am that which I must do.

Much of our contemporary difficulty with pleasure comes from the fact that the relations between work and leisure have become so blurred and uncertain. The response we should make is not to try to clarify or reassert the difference, but to enter into it. Because we are not as sensitive as we might be to the complex economies of work and play, our reasoning about the kinds of reasoning that are at work all the time in our experiences of work and leisure, and the pleasure that runs back and forth between them, is often fuzzy and feeble, and so not nearly enough fun.

For perhaps the thing that gives us the greatest difficulty with pleasure is that pleasure has no obvious or permanent contrary. This is because of the many ways in which pleasure proves itself able to get on the other side of itself, to inhabit and turn to its own account the many things that seem lethal to it, even and especially, as Freud shows in *Beyond the Pleasure Principle*, death itself. It is not possible to be any kind of rational human being without some complement of masochism, the deriving of pleasure from pain or, perhaps more precisely, the intensification of pleasure through it. Human beings have been characterized as the only species that voluntarily eats chilli peppers – *Homo capsaicus*.

If we have difficulty with pleasure, it may come from the increasing abundance and availability of pleasures, forcing more and more people to internalize the limits and forms of regulation that scarcity had previously provided, as well as from the formalization of pleasure that is a concomitant feature of the increasing abundance. This has produced a kind of Romantic blur and blunder about the ways in which pleasure is in fact intertwined with number and measure. Rather than attempting to rescue or purify pleasure from the difficulties it has got into, I have wanted to emphasize the paradoxical fact that pleasure is in fact entangled with, and even in some sense dependent upon, difficulty. If we have difficulty with pleasure, this is in part because we seem constituted to get such pleasure from difficulty. Rather than an idealizing or essentializing quarantining of absolute pleasure, I recommend an enhanced utilitarianism, which cleaves to the principle that only utility can determine value, but also recognizes that there is no single unified currency, or principle of mensuration, by which pleasures can be totted up, even as pleasure is, *ab initio*, and ever more irreducibly as time proceeds, utterly suffused by quantity and number. Leibniz was right, though not for reasons of which he is likely to have approved. If we are to understand and account for our pleasure, then measure is indispensable. *Calculemus.*

7

PLAYING THE NUMBERS

One of the most important of the ways in which we have come to live in and through numbers is through the heightened awareness of risk and probability that characterizes modern experience. All human cultures are intensely aware of the risks and dangers to which they are exposed, and concerned to anticipate and mitigate them as much as possible. One might even define a culture as the effort to reduce unpredictability, to create what is in a state of nature highly improbable, namely a maximum of predictability. Many civilizations have counterposed ordering functions like law, custom and religious ritual to the terrifying effects of unruly or chance events. But most have done so in terms of the contrast between absolute law and completely unpredictable chaos – destiny, as it were, and fickle fortune. All that changed in the course of the seventeenth century. Suddenly, chance seemed to become calculable. Chance has become more and more a matter of specific probabilities, expressed above all in numerical ratios. As chance has progressively become a matter of number, so probability has become an important and powerful branch of mathematics. Where number had previously been identified with the definite, it now moves between, and itself mobilizes, the definite and the indefinite.

This chapter concerns itself with the challenge to the understanding of artworks represented by the numbers games of probability. It considers first of all the changing nature of literary texts, and then examines the pressure that the mathematics of probability have been putting on art and art criticism from the late nineteenth century onwards.

When I read a book, what are my chances, or, as one might as well say, what are its chances, that I will get to the end of it? If I do, how likely am I to have read it all with optimal attentiveness? Will I have been able to read it concentratedly, all in one stretch, or will I have been plagued (or relieved) by interruptions? If I do finish it, will I ever return to it, and, if so, to reread all or merely part? None of these factors ever features in my thinking about or teaching of literary texts, though they may sometimes be mentioned when discussing study skills (always try to read in a good light, away from distractions, and with a pencil in your hand). The text is hitched, in a marriage made in the heaven of readerly and critical conception, not only with its ideal, predestined reader, but with its ideal prescribed reading. If we do not finish a book, it is we who have come up short, not it. It is not so much that we do not attend to these matters, as that we regard them as in principle neither worth attending to nor in practical terms the kind of thing of which it would be possible to take account. They are purely contingent factors.

To be sure, 'the reader' does make occasional appearances in literary theory and criticism, as does 'the viewer' in discussions of art, though my focus in most of this chapter will be on readers of texts rather than works of art more generally. This reader is said to be 'situated' in various ways, which means that they can be assumed to come at the text from various predictable or predetermined angles. Sometimes, this reader is said to be 'plural', unresolved, resistant, refracted, refractory or cross-grained. But, in order to be spoken and written about at all, in the manner in which it appears such things are required to be written and spoken about, the reader has to be brought over from the side of contingency to that of necessity. The contingency of a reading is construed as a determined, predictable or necessary contingency. But the point about contingency is precisely that it is not fully predictable. It is not only a contingent matter what sort of reader or viewer I am (most critical writers seem able to distinguish only a small number of these sorts – gender, ethnic affiliation, class, sexuality and degree of disability just about cover it), but a fully (which of course is always also to say, partly) contingent matter how far I will on any particular occasion in fact conform to these determinations.

Just occasionally, rumours of this contingency can penetrate to the interior of the literary text, as in addresses to the reader of various kinds, or in the permission that artists will often magnanimously grant to viewers to make of their work what they will (thanks for that, but I was going to anyway). The form of literary fiction might partly be determined by its effort both to acknowledge and to head off the liability to interruption that long texts are almost by definition heir to, not by attempting to keep the reader grimly glued to the reading from beginning to bitter end, but by conceding the strong likelihood of interruption and encouraging the reader to synchronize his interruptions with those provided by the text, in a kind of preemptive choreographing of contingency.

On the whole, however, literary criticism acts almost entirely as though it were functioning in the domain of law and necessity. In fact we might define the concept of a text, or The Text, as the rendering of the contingency of reading as a necessity. This might well, I think, strike us as odd, given that most of us would regard the investigation of reading and writing as much closer to the grain and fluctuations of things than, say, the pursuit of mathematics, or the measurement of air pressures. Not only this, literature itself seems to take as its subject, not the sphere of necessity but what Thomas Hardy calls 'change and chancefulness'.[1]

Those who have written about the relation between literature and probability have tended to do so in terms of the ways in which probability features in it, or of its ideal reader's response to it. In Robert Newsom's *A Likely Story: Probability and Play in Fiction* (1988), for example, the word 'plausibility' might be substituted throughout for 'probability', since it is concerned almost exclusively with the ways in which fiction represents conditions of doubt or uncertainty, mobilizing forms of probabilistic judgement in its readers.[2] Its concern is therefore to explicate the effects of a background of ideas about probability on literary writing, and the ways in which those perspectives then feature within that writing, especially the fictional realism of the seventeenth century onwards. The questions asked concern the judgements that can be made about the likelihood, lifelikeness or convincingness of the actions of characters or depictions of worlds in literature. But all of this occurs within the dubiously determined and the determinately dubious space of the literary text. The concern with the ways in which probability is deployed by literary texts leaves no space for the way in which literary texts might themselves be exposed to conditions of chance, or themselves operate within fields of probability.

PLAYING LITERATURE

The formalized estimation of probability has been condemned by Nassim Nicholas Taleb as relying on and promoting what he calls the 'ludic fallacy', the idea that events in the world are best understood by formalizing them as games, which is to say with a 'flat' background of equal chances.[3] Under these circumstances it becomes possible to solve classical problems like that of how to allocate the winnings fairly in an interrupted game of chance, on which both Galileo and Pascal cut their probabilistic teeth, thereby inaugurating the mathematics of probability. But what might it mean to think of the work of art or literary text as a game?

There is in fact a substantial history of associating art and play, literature and game, as well as a slightly less substantial literature in which game or aleatory procedure is involved in the generation of artworks themselves. But, though art and literature can explore or even incorporate gamelike structures and procedures, a game is never an exposure to the open as such, for there is no open as such. Indeed, games seem in important respects opposed to pure contingency. All games are in fact determinate generators of indeterminacy. The rules of a game may attempt to cover every contingency, but they can never predict or exhaust it.

This is perhaps imaged in the astragalus, the animal heelbone which was the favoured form of randomizer for a very long period among Egyptians, Greeks, Romans and other peoples. The physical form of the astragalus is a graphic allegory, or what is called a 'phase portrait', of the blending of the determined and the undetermined in the game that is played with it. The astragalus can land on any one of its four faces, but, since there is no standardized form of the astragalus, the chances are not evenly distributed between these two faces. It is, as we say, weighted or biased in different ways. In the

Jean-Baptiste-Siméon Chardin, *The Game of Knucklebones (Les Osselets)*, 1734.

classical world, these faces counted for one, three, four and six, the numbers two and five being omitted. There seems to have been about a 10 per cent chance of throwing a one or six, and about a 40 per cent chance of throwing a three or four.[4]

Every astragalus has two bodies, an actual and a virtual. There is first of all the bone itself, in the awkward aggregate of its angles and oddities, the lumpy three-dimensional landscape of likelihood, that is both given and yet unknowable, or as yet unknown, wholly apparent, yet entirely unpredictable. Play begins. Imagine please that there is one who is recording the sequence of throws as they are called out who cannot see the game and has no knowledge of the shape of the astragalus. For a long time, he can tell little or nothing of the shape of the astragalus from the sequence of throws. But slowly, even inexorably, over time, and after many, many throws, assuming the willingness or capacity to keep perfect records, the ghost of the astragalus's shape, the abstract law of its distribution of possibility, may begin to emerge. Putting the astragalus in play will expose its physical form to randomness, which will initially scatter that physical form into indistinctness, giving the astragalus something like the perverse shape of contingency itself. But the pure contingency that at first scrambles the shape of the astragalus slowly starts to reassemble its form, albeit now translated into a kind of numerical distribution, a little as the scribblings of the crayon reveal an otherwise indistinct form in a brass-rubbing. Eventually, one's data on outcomes will start to come together in a kind of virtual astragalus, a distribution of probabilities that will be the stochastic silhouette of the original.

It might seem at first as though the rather ungainly shape of the astragalus would make it harder to guess its shape, but in fact this will tend to make it easier. The uniform haze or blizzard of randomness will make the oddity and unpredictability of its knobs, ridges and declivities stand out more clearly than that of a more regular shape, just as a word is easier to guess when one has only consonants as opposed to vowels, since consonants are less common than vowels – one uses the abbreviation 'rptn' for the word 'representation', not 'eeio'. A shape that is closer to equilibrium, which is itself more likely to generate random or unpatterned results, will keep its head down much longer in the hail of circumstance. A roulette wheel, or a ball, may escape detection for very much longer than a coin or die. You may recognize in what I have been describing something like the process involved in guessing the nature of the Enigma machine by the codebreakers at Bletchley, who were faced with the problem of inductively determining the physical construction of an encoding machine that was designed to produce randomly scrambled outputs using only those outputs themselves – by indirections finding direction out. In both cases, the sheer mass of random outputs allows a slow building of a determinate shape, instrument or process. Something like this process has also been put to work in modelling procedures – for example the simulation of biological processes like the human immune system.

A game is a putting into play, in an attempt to model this emergence of necessity from contingency. But where the conditions of a simple game are given in advance, and its possible outcomes limited, there are many kinds of game situation in which what is being sought through the play is not just the shape of a particular object that the game puts into play, but the shape of the game itself. I think that a more quantitative outlook on interpretable artefacts may make it ever more interesting to think of reading literary texts, or, for that matter, watching a film or listening to music, as just this kind of playing of a game, where the nature of the game itself is only semi-determined, or itself must emerge stochastically from its playing. We are accustomed to a much simpler kind of model of literary texts and their reading. On the one hand, there is the text, which is a given; on the other, there are its readers and their readings.

In the case of a literary text, what one is attempting to model through the trial and error of reading-play is the act of modelling which is reading itself. It is as though the challenge now were not to use the distribution of outcomes to model the astragalus alone but also to model the conditions under which the game is being played – the kind of surface on which the astragalus is being thrown (a sandy floor? a table with a tilt? a counterpane? a tray in a chariot being drawn by two skittish chestnut yearlings?), the number of players there are, and their idiosyncratic throwing styles (how high does each player throw the astragalus? how far does it usually roll?). In the first case, there is complete information, and the possibility of a complete mapping of the probabilities; in the second, the information is incomplete, and the judgements necessarily inductive.

In every game, there are perhaps two contrary motivations that the playing of the game itself ties together. The first is the desire to create conditions of randomness. The second is to use those conditions of randomness to disclose the game's own essential form. A game always asks the question 'What kind of game am I? What is possible within my limits? How much play do I allow and afford?', inviting the bringing of necessity out of contingency. The point of playing a game is not only to win at it, but to figure out in the process what sort of game it is. And, we will see, this question is always asked in a looped future perfect tense, or what is called in French the future anterior: thus 'What kind of game will I turn out to have been?' 'What kind of play will have been afforded by the kind of game I will have been revealed to be?' This doubleness is indicated by the fact that we use the word 'game' both for the set of rules and procedures that constitute a game (the game of chess) and a particular episode of playing, or actualization of the possibilities of the game (a game of chess). The two meanings of game, those signified by the definite and indefinite articles (the game of chess and a game of chess), are always both in play.

SEQUENCE AND ENSEMBLE

One of the greatest difficulties in making sense probabilistically of literary texts, which might be thought of in some ways as singular historical events, is the incommensurability of sequence and ensemble in probability theory. Put more simply, this is the principle that probabilities only apply to large collections of events, and can only measure the likelihood of certain events occurring over the long term, as a result of repeated trials, never in the short term, or at a particular point in a sequence. Probabilities predict what happens on large scales and cumulatively in collections of events. They tell us nothing about the order in which those events are likely to happen. Toss a coin 500 times, and, even if it comes up heads 499 times in a row, there is still a 50–50 chance of it coming up heads again on the 500th toss. If there is a likelihood of a particular sequence of numbers being drawn twice in the space of a hundred years, this tells us nothing about where in the sequence those two numbers are likely to occur – which means that they are just as likely to appear next to each other in successive draws as they are at fifty-year intervals.

Our relation to information in particular is more and more a relation to numbers like this. It is difficult for us to know how to be of one mind about these numbers, because they slice our existence in two. Living in a world of abundant information means I have to be simultaneously in the world of big numbers (ensembles), and the world of small numbers (events), of which I myself am one, indeed the very form that my idea of 'one' takes. The most important kinds of information to which I have access and on which I feel I ought to be able to act rationally tend nowadays to come in the form of big numbers – the numbers of people succumbing to stroke, owning cats, burning coal, voting Conservative or simply existing in the world – and since you read the phrase 'big numbers' a couple of seconds ago, unless a Person from Porlock has intervened, the number of people in the world has gone up by 47 (actually closer to 48, but even numbers sound much more rough and ready than odd). Big numbers can, I know, in the long term, be counted on. On the small scale, by contrast, in the world of one thing at a time and one thing or another in which, as a finite creature skewered in space and time, I have to live, things are sputteringly, spasmodically erratic. For example, on average, global road traffic fatalities seem to chug along at a rate of about one every 25 seconds, with suicides limping slightly behind this figure, at around two a minute. But this gives little help in knowing what my actual chances are as I am poised to cross the Euston Road. We can assume that, like buses, traffic fatalities come along in clusters rather than on a regular schedule, with peaceful lulls in the global death-count for minutes at a time, followed by spectacular, screeching pile-ups which increase the tally by a dozen or so at a stroke. Similarly, hovering spoon in hand over my triple-decker Death by Chocolate, I may reflect on my personal cardiac odds. I may have a 10 per cent chance of a heart attack in the next five years, but I have a 0 per cent chance of having 10 per cent of a heart attack. Nor can I smooth out the risk by putting it on a sort of existential tick, since I also have a 0 per cent chance of having 2 per cent of a heart attack in each of those five years. No, I'm either in for a heart attack, a whole and juicily irrefutable infarct, or I'm not. The more information I have, the more I am split between these two forms of accounting, one to the right and the other to the left of the decimal point, in one of which I have a 10 per cent chance, and in the other of which, all along, my chances will only ever have been 0 per cent or 100 per cent. I am scissored between these two worlds, both of which indubitably exist, and exist inseparably from each other, since, after all, the big numbers are just all the little ones added together, yet I cannot live, or at least cannot turn out to have lived, in both at the same time. There is either safety in numbers, or my number is up. We may recognize in this scission the intersecting orders of numerology and numerality that have been evoked a number of times already in this book.

Since probabilities relate to ensembles and not to entities, this may seem to imply that probability considerations have no purchase on the individual items we know as literary texts. But there is one important sense in which the texts we know and denominate as literary might seem to qualify amply as probabilistic ensembles, namely in the fact that, by definition, literary texts tend to be experienced more than once. Putting it at its simplest, literary texts are texts with a higher than average probability of being reread. The phenomenology of rereading may not appear much in accounts of literary texts, but it is its implicit condition. To consider a text a literary text is to suggest that it requires or is liable to rereading, either locally, sentence by sentence, or paragraph by paragraph, or globally. To say that a text requires or is susceptible to rereading in these ways is to say that there is a higher probability of this rereading than with other texts, not because they are already literary texts, but because literary texts just are texts that happen to be subject to this higher probability of pouring rather than poring. We might even see Roland Barthes' specification that 'Literature is what gets taught' (*ce que s'enseigne*) as another way of saying that literature is what gets to be reread.[5] On this specification, a literary text is therefore a text that has a high chance of being treated as a literary text. Literary reading is sometimes defined as the bringing to bear of a certain kind of attentiveness, one that is attuned to questions of linguistic form, for example, but it really involves any kind of unnecessary or surplus reading, and the features that it may disclose. And, even if one sets aside the fact or horizon of individual rereading, the fact that literary texts are texts that are studied and discussed makes them texts that are experienced as multiples, as the aggregation (and exchange) of serial acts of reading. We think of nonliterary texts as much more likely than literary texts to be characterized by average or predictable kinds of response, but in fact nonliterary texts are much more likely to be experienced uniquely – that is, once and for all, not in the horizon of predictable alternative readings. From this point of view, it is in fact literary texts that are read by means of the averaging of divergent responses, and by the diverging of average responses.

For this reason, literary texts ought to be much more responsive than other kinds of texts to the model of reading as a series of iterated plays, occasions or chances of reading, which in turn makes it apter than it otherwise would be to think of them as ensembles rather than singular entities or events. This would make the text entitled *Middlemarch*, for instance, something like the rules of a game of which each reading is the enactment. As one plays the game called *Macbeth* or *Aurora Leigh*, one builds up a map of its probabilistic landscape, like the one who tries to intuit the shape of the astragalus or Enigma machine from the sequence of plays to which it gives rise. The probabilistic figure that the text cuts, or the stochastic landscape it seems to delineate, is both actual and virtual; it is never all together in one place and time, but only ever a set of possibilities, even though the distribution or physiology of those possibilities may start to seem indubitable.

And, remembering the analysis ventured earlier, we need to add this. In reading *Macbeth* and *Middlemarch*, we are not just guessing at the kind of plaything or chance-dealing instrument it is: we are also guessing at the kind of game we are playing, since there is in this case no text that really stands outside or before the game begins. One does not in fact simply put the text into play, since the text is the outcome or profile of this putting of it into play. The goal, or at least the process, of a game of *Macbeth* is to disclose the shape and reach of the game called *Macbeth*. The practice of literary criticism that forms the background of all the kinds of reading we might think of as literary is increasingly an agonistic one, in which to read the text is to decide whether to accede to readings of the text offered by others or to develop my own, which is what winning at *Macbeth* might mean.

As one rereads, one encounters and foregrounds the relations between redundancy, or features of the text with high probability, and information, namely unpredictability, or features of the text with low probability. Redundancy is used here not in its everyday sense of uselessness or unnecessariness, but in the sense employed by information theorists, who indicate with it a certain quota of excess or repetitiousness. The redundancy of a message is the amount of information required to transmit the message minus the amount of information needed for the message itself. Every utterance involves elements that are not necessary to the specific utterance, elements that simply register or confirm the fact of the utterance taking place, or indicate the structure of language. The word redundancy, which derives from *re* + *undare*, to come back in waves, can also mean echoing or resounding, which aptly suggests the role of redundancy in turning the message back on itself, the channel checking that there is contact, which is to say, that there is, that it is, a channel, saying yes, this is a message, are you on the line, are you still receiving me, do you get it? Without this apparent excess, no message can in fact be transmitted. In a sense, redundancy can be identified with the channel or form of the message, which must involve recognizable, repeatable elements.

No text or message can consist of either redundancy or information exclusively, and neither redundancy nor information can exist independently of the other. There will be features and procedures that become familiar in the text, and there will also be features and procedures that we will recognize from the reading of other texts. These are not given in advance, for our recognition of them will itself be a contingent matter, which depends upon a number of variable factors, most importantly how many times we have read the text in question before or how many other texts one may have read.

These redundant or high-probability features will tend to predominate and in certain cases may end up by inundating and therefore exhausting the text by starving it of information, by which I mean the emergence of low predictability out of a background of high probability (we will see soon that, since fields of probability are not static distributions of value, but as dynamic as weather systems, a state of high probability can sometimes begin to make the appearance of low-probability events more probable).

Of course texts do not merely enter the condition of reread or rereadable texts by fiat. We may perhaps say that all texts wish in some sense to continue in their being, by which one means, not to endure exactly, but rather to continue to be subject to replication. I don't mean this literally, though there are no doubt some features of literary texts and their readings which are partly determined by the conscious interest of their writers and readers in perpetuating them, which is to say, converting them from singular into serial entities. In fact, the wish to remain in being is not to be thought of as programming and impelling the text, but instead as emerging from the tendency of certain texts in fact to remain in play, or get themselves reread, as a result of their conformability to changeable sets of conditions. The will to remain in being of a given text is therefore in fact the probability of its successively doing so. A plant the leaves of which grow round its stem in a ratio approximating to that of the Golden Ratio is not deploying this strategy or striving towards this form as a way of maximizing its chances of survival, though putting out leaves at these intervals does maximize the amount of sunlight it can gather and also give it the greatest chance of shutting out the light from competing plants beneath it. Its will to obliterate its competitors is a retroactive artefact of the fact that it turns out that it stands a good chance of doing so.

Whether or not a text gets to meet the condition of becoming thought of as literature, which is to say, a text that will be reread, is itself a matter, not of 'pure chance', a notion that will need a little later to be put under some pressure, but certainly of unpredictably variable probabilities. Some texts will achieve rereadability, some will have it thrust upon them, and some will not – and whether or not they do achieve it is correlated hardly at all with whether they aspire to. In fact, though, the idea of continuing in being must be thought of, like almost everything else in this kind of evolutionary perspective, back to front; that is, not as an engine shoving things forward from behind, but as a property or propensity that gathers over the course of time, and as a result of replication. The will to persistence of writing is a back-formation of the fact of its inertia, or tendency to persistence. The disposition or capacity to replicate is built into all writing, since it can be read many more times than it can be written, unlike speech which, until it can be recorded, that is, until it can become writing, can only be heard as many times as it can be uttered. Something that happens to survive starts irresistibly to take on the appearance of meaning or having been meant to. Why do some texts persist for longer than others? Because it turns out they can. Given the in-principle replicability of all writing, and given also a differential and temporally changeable landscape of readerly habits, motives and preferences, the chances of all texts lasting the same amount of time, that is, of there not being texts that last longer than others, are vanishingly small. Let us not forget the fact that, in reading, as in the expanding population of grey squirrels or Japanese knotweed, nothing succeeds like success.

To say that literary texts are ones that are subject to a high probability of being reread is to say that they are texts that have a more extended temporal profile, which is more than saying that they simply last longer than other texts – for it is not a question of simply persisting, so much as radiating. Radiation does not mean dissemination necessarily; texts do not always decay into proliferation or polyvocality – they sometimes decay into univocality. In order to persist as rereadable entities, literary texts need to provoke and survive significantly variant readings, to consent to seem difficult, in agreeable ways, to add up.

LASTING

It is certainly true that we should not expect to be able to quantify exactly the fields of probability which condition the chances of survival and propagation for cultural artefacts and ideas. On the other hand, we will not be able either to dispense entirely with the notion of quantity, simply because the qualities that we make out in literary texts will in the end come down to numbers and frequencies, even if they are only specifiable in terms of statistical averages, ratios, estimates and rates of change rather than precise calculations. But this is in any case the nature of statistical analysis, which offers a way of calculating on relative rather than exact quantities, a way of getting as good a fix on imprecision as one can. One example of a kind of quantitative analysis is that offered in Franco Moretti's *Graphs, Maps, Trees* (2007). Moretti proposes and explores a mode of 'distant reading' that would use long-range quantitative evidence, rather than the microscopic reading of individual texts, to show the development over time of what he calls a '*comparative morphology* of form'.[6] Though he does not address questions of probability directly, they are the engine of most of the effects that he analyses. His first chapter, 'Graphs', maps the distributions and periodicities of fictional genres. The evidence that Moretti presents suggests that the majority of fictional genres – the silver-fork novel, the Newgate novel, Imperial Gothic – appear to flourish for around twenty to thirty years, and then rapidly and all at once give way to others (*Graphs, Maps, Trees*, 18–19). This suggests that genres are related more than etymologically to generations, in that their lifetimes are synchronized. Moretti's proposal is that, though one can sometimes suggest external triggers for the birth and supersession of genres, generic generations are in fact internally paced, simply by the rhythm of a group of individuals who are drawn into solidarity by a particular destabilizing prompt, then persist by a kind of inertia until their solidarity with each other begins to weaken:

> Once biological age pushes this generation to the periphery of the cultural system, there is suddenly room for a new generation, which comes into being simply *because it can*, destabilization or not, and so on, and on. A regular series would thus emerge even without a 'trigger action' *for each new generation*: once the generational clock has been set in motion, it will run its course – for some time at least. (*Graphs, Maps, Trees*, 22 n.11)

Moretti places his analysis in a medium term that lies between the micro-time of individual events (particular literary texts, for example) and the macro-time of the Braudelian *longue durée*. This time is populated by cycles, for 'the short span is all flow and no structure, the *longue durée* all structure and no flow, and cycles are the – unstable – border country between them' (*Graphs, Maps, Trees*, 14). We might perhaps, without impoliteness, rewrite structure and flow as redundancy and information. These temporary structures – 'morphological arrangements that *last* in time, but always only for *some* time' – are formed of probabilities (*Graphs, Maps, Trees*, 14). To say they last is to say that they recur, which in turn is to say that they introduce islands of redundancy or high predictability into fields of relative disorder or unpredictability.

As Moretti makes clear, the advantages of this method depend upon the availability and the manipulability of evidence that in turn enable us to shift scale in the required way. This does not just involve the massing together of the individual units we call literary texts into larger aggregates that allow us to measure distributions. For it is possible also to try to make sense of the patterns of distribution across these aggregates of features that are smaller than texts – in the cases that Moretti offers, the 'clue' in detective stories, or the device of free indirect style. This makes it possible for him to say that the forms that shape literary history are simultaneously 'the very small and the very large', the motif or, as a structuralist of a certain stripe might once have called it, the lexeme, and the corpus (*Graphs, Maps, Trees*, 76). A quantitatively based literary history of the kind that Moretti proposes would use the former to generate the latter, with texts being the carrier-form that drops out of the picture. The analogy might be with the analysis of the distribution of genes and alleles in different populations, in which the individual bodies that are the bearers of these genes similarly fade from view. 'Texts are certainly the *real objects* of literature . . . but they are not the right *objects of knowledge* for literary history,' Moretti claims (*Graphs, Maps, Trees*, 76). He may mean by this in part that texts may not provide the best samples from which to generate large amounts of data.

It is, I think, a promising strategy, which suggests that finding the right kinds of molecular elements within texts, or other objects of critical and historical attention, might allow for the determinate measurement of molar fluctuations. Given that the unit of currency of most of the non-mathematical databases in the world today is the word, for example, we ought to be able to devise a calculus of word appearances and meanings. The probabilistic spectrum represented by a single word-entry in the *Oxford English Dictionary* suggests that it might be possible to begin to base our judgements on the meanings and functions of words at particular times on estimations and inferences of probability rather than the crude averaging and rounding up or vague notions of drift and transition on which we currently rely. Instead of asking what a word meant in 1594, we might ask what its chances were of being understood in a particular way. The implications for larger structures, like cultural or literary historical periods or movements, which are themselves also roundings-up (often relying on hair-raisingly unrepresentative samples, with a correspondingly huge likelihood of error) yet powerfully determine the ways in which we legitimate our knowledge and judgements, are immense and mouth-watering.

There will for some time continue to be objections that this is no more than a new round of positivism that looks to what are fantasized as 'the sciences' for a spurious and inappropriate exactitude. In fact, I think it may be the opposite – namely, a way for the humanities to escape the intractable positivism that in fact lurks beneath its convictions of the approximate and the indeterminable. In this, the humanities may in fact be borrowing back something they in the first place lent to the sciences. The 'social physics' that is adumbrated by writers such as Philip Ball is an interesting recall of the positivism of Quetelet and others who first began to apply statistical methods and reasoning to the understanding of social phenomena.[7] There was certainly a deal of overconfidence in those who thought prematurely that they had discovered the invariant laws of human and social behaviour. But it seems highly probable that the social physics of the early nineteenth century had a decisive influence on scientists like James Clerk Maxwell, who were faced with the problem of calculating the behaviour of physical substances, like gases, in which it was impossible, practically and in principle, to account for the movements of every single particle, and who drew on statistical thinking in the new social sciences to develop stochastic models in the physics of matter. When Einstein used probability theory to explain the mysteriously erratic dance of pollen granules noticed by Robert Brown in 1827, and known thereafter as Brownian motion, he decisively established the importance of statistical physics.[8]

If we accept, as I think we should, that the study of literature in its historical context is in fact a study of mass phenomena, and requires the generation of inferences about very large ensembles of similar and recurring events (a book being read, a symphony being heard or a picture being contemplated), we should not be indifferent to the ways in which large ensembles of phenomena have been analysed in other areas, in order to try to make our models and methods less bungling and dubious. I would like it if we tried to find many more things to count and measure and many more ways of counting and measuring them rather than develop ever more sophisticated theoretical models that are based upon the most primitive kind of knock-on, cause-and-effect physics. For the humanities, especially the theoretical humanities, that aim to model the processes whereby cultures and subjects are formed, are locked into a determinism that is date-stamped about 1750 (though with little responsiveness to the major advances in probability theory that had already taken place by that time). The more empirical and historical forms of the humanities are less arrogant but scarcely less deterministic in their understanding of causes and relations. All fail miserably at any predictive test of their competence and value. Everywhere, we are asked to believe in the existence of the simplest, most remorselessly linear processes which act evenly, uniformly and predictably. Interactions of any complexity at all are almost entirely absent from this writing. Nearly all of these explanations depend upon staggeringly naive faith in the adequacy of the skimpiest and most schematic accounts of initial conditions to explain outcomes; everywhere, that is, there is a dependence on what Daniel Dennett calls the 'mind first' fallacy, the idea that forms can only emerge from prior models, and that nothing in what emerges can not have been latent in what it emerged from.[9] The humanities have surprisingly little tolerance for error, exception, anomaly and emergence, for things that form from unpredicted and probably unpredictable conjunctures of circumstances. The only models that count in the humanities are determinist models in which what happens can only ever be the actualization of specific and knowable potentials.

CHANCE WOULD BE A FINE THING

Nowhere is this more, or more lamentably, apparent than in the ways in which the topics of chance and indeterminacy themselves appear in writing about art and literature. It is perhaps not entirely surprising that the operation of chance should be presented as the absolute and incalculable Other of law and determination in this way. The force of 'chance' is fetishized as a power that is alien to every kind of determination, rather than being threaded through it. Tristan

Tzara's well-known recipe for making a chance poem may help us understand this:

> To Make a Dadaist Poem
>
> Take a newspaper.
> Take some scissors.
> Choose from this paper an article of the length you plan
> your poem to be.
> Cut the article out.
> Now carefully cut out each of the words that make up
> this article and put them in a bag.
> Shake gently.
> Take out each cutting one after another, in the order in
> which they left the bag. [sic – but the second clause
> of this sentence should clearly come after the two words
> in the following line]
> Copy conscientiously.
> The poem will resemble you.
> And there you will be – an infinitely original author
> of charming sensibility, even though unappreciated
> by the crowd.[10]

The most immediately striking feature of this text is how much determination is threaded through its chance operations. To begin with, one must decide, or, rather, have already decided, how long the poem one wants to write must be. Indeed, prior to that decision, one must already have decided to write a poem, and a Dadaist poem at that. One must choose one article, and take care to cut round each of the words in the article. The bag in which the words are to be reordered is to be shaken 'gently' – as though it might invalidate the result, or skew its chanciness in some way, to agitate it too vigorously, perhaps by shaking it back into orderliness. One must copy the words out 'conscientiously', and must preserve the order in which they came out of the bag.

The result is that 'The poem will resemble you.' Is this supposed to be the triumphant consummation of the aleatory operation, or its hapless defeat? Perhaps Tzara is suggesting that the vigilant suspension of everything that might exert conscious influence over the operation will give access to the unconscious essence of the person doing the selection (this being a common promise made about chance operations in Surrealism and Dada). But perhaps he is simply pointing to the fact that, given the strict armature of the aleatory ritual, the resulting poem cannot help but be taken as an expression of the aleator. Perhaps Tzara's words mean the words thus arbitrarily assembled will, *mirabile dictu*, themselves provide a revelation of the hidden reality of the assembler, which will have been liberated by the process. But perhaps he might also mean that the miraculous benediction of the poem will be the very opposite of a chance occurrence, precisely because it will have been so carefully and rigorously set up, and because the magical procedure makes it exceedingly likely that, whatever the result will be, it will be bound, like a horoscope, to seem like some spooky miracle of aptness.

I think that Tzara's recipe, which is usually quoted as though it were itself a poem, though, if so, it would seem that it could not, by its own design specifications, be a Dadaist poem, is readable (or is now) as a wise reflection on the trickiness of achieving chance. Pure chance can only be guaranteed by strict determination, because 'chance' cannot be relied upon to happen by chance. So 'mere' chance has a good chance of being impure, contaminated by determination, in this case the predisposition of anybody likely to engage in this divinatory procedure towards making out significance in what results. Chance, like death, is hard to avoid, until one resolves to embrace it, at which point, like death again, it becomes elusive. Furthermore, the recipe seems to recognize that chance does not stay chance. Like Lady Bracknell's ignorance, chance 'resembles a delicate exotic fruit; touch it and the bloom is gone'.[11] The recipe intimates how difficult it is to cross over entirely on to the side of chance; seemingly, it is as hard to get chance into one's poem as it is to keep it out.

Dadaism was only one of the areas of art practice to become interested in trying to exploit the operations of chance. This is something different from simply playing the odds, in the way in which a gambler might, since a gambler only wins if he is lucky. The sort of betting on chance engaged in by Dada is of a kind that, as long as the chance procedure is constructed carefully enough and the mechanisms of the aleatory procedure followed to the letter, the player of the game cannot help but get lucky, since they will always and without fail be exposed to the operations of pure chance.

The awareness of 'chance' has led to a tendency to reify it. As well as becoming a substantive, chance has tended to become an adjectival quality, as in the contemporary use of the word 'random', or in its absurdly intensified form 'really random'. Here 'random' means something like 'pleasantly unexpected' or 'quirkily unpredictable'. But randomness has no specific quality, no defining tone, hue or cast, for randomness is the absence of any determination whatever.

The word 'chance' does not signal a force of pure randomness. We have a tendency to think of chance as a kind of loosening or dissipation that scatters coherence and breaks open regularity. But chance is not all on the side of incoherence – if it were it could not have given rise to coherence-creating species like human beings. For, where habits and regularities form, these are not opposed to chance, but themselves arise from it. Regularity may be much less probable than irregularity, but it is not in any sense opposed to or on the other side from chance. To say that something happens 'by chance' suggests something that happens without a cause or reason. But there is no simple division between things that seem to happen for a clearly determinable reason, and those that do not. For most of the things that do have a cause or reason, those causes or reasons are not absolutely determining, and whether or not they are determining is itself contingent. Whether one thing will turn out to be dependent on another is usually itself dependent on other things still. Only causes or reasons that are absolutely necessary and not just highly probable are immune from chance – but there are very few, if any, of these. There is, as the devout James Clerk Maxwell was reluctantly obliged to accept, no absolute law decreeing that closed energy systems move from a condition of lower to a condition of higher entropy, that is to say, a condition in which, though exactly the same amount of energy remains in the system, it is distributed in such a way that less of it will be able to be converted into work. That this will tend to occur is not necessary, just very, very probable. The Second Law of

Thermodynamics is therefore no such thing, just a racing certainty. Every time entropy does in fact increase, it is dependent on the chance that, once again, the more probable rather than the less will have happened, when it did not absolutely have to. The chances may be a squazillion to one that the molecules in a cup of hot coffee will one day all spontaneously line themselves up into a neat holographic portrait of Princess Diana shimmering in mid-air, but, since there is nothing to prohibit it absolutely, every time it does not happen, it is a matter of chance rather than necessity that it has once again let us down.

This complexity may put a new complexion on the interest in the operations of chance that arose in many different art forms at the beginning of the twentieth century. In an obvious sense, this interest looks like both recognition of the unavoidability of chance and an acknowledgement of its generative powers. Artists of many different kinds have seen an openness to chance as one of the most powerful forms of resistance, discovery and renewal in a world characterized increasingly by rational management, and the apparent reduction of every kind of risk. Others have followed spiritualists, themselves following divinatory history, in seeing it as a path into knowledge of the unconscious, chance being allegedly a way of outwitting the sentries and censors of rationality. These powers of resistance and renewal depend upon a conception of chance as a kind of pure exteriority to reason, or to the reasoning subject. But chance is in fact never available as this kind of absolute exteriority, or in any sort of 'pure' form. The art that would make chance an exterior force on which to feed will always be liable to encounter the force of chance as part of its own operations, and intertwined with its most deliberated purposes.

This is because it is always possible, by chance, that some disappointingly or suspiciously orderly arrangement might arise in any undetermined procedure. None of us would be very convinced if, in response to the request to provide a sequence of six numbers at random, a program were to generate the sequence 123456, but there is quite a significant chance of such a sequence arising at random. As a sometime historian of and speculator on the voice, I have had occasion to enjoy and endure a number of episodes or performances of glossolalia, both in artistic and religious contexts, in which sounds are emitted that are said to be pure nonsemantic utterance, or at least to belong to no recognizable earthly language. The interesting feature of such utterances is that, far from being impelled by the pure language of the spirit, or of the elemental passions, they always in fact seem to be subject to careful internal monitoring, so as to avoid the accidental articulation of meaningful words. Given that many of these words arise from the crystallization of accident out of the mouths of babes and sucklings in many different times and climes, it is highly improbable that an entirely unfiltered stream of spontaneous utterance would not occasionally contain them, yet I have never heard a glossolalic performer come near to articulating 'mummy' or 'pop' or 'bugger' or 'haddock'. In order to count as entirely open, such speech cannot in fact be open to simply anything and everything. The order of accident must be tacitly defended against the accident of order.

Seen in these terms, the ideology of chance may be seen as the effort to disavow this intermingling of the determinate and the indeterminate – an intermingling that can never itself be fully determinate or calculable, though this does not make it incalculable either. What we may call the aleator, or artist of chance, is therefore the mirror image of the determinist; where the latter strives to leave nothing to chance, the former is at pains to have absolutely nothing go to plan (except that).

Works on the operations of chance in different art forms tend to focus on the ways in which such forms might or might not succeed in surrendering or opening up to a principle that is held to be alien or antagonistic to its nature. Such a perspective allows one to rest safe in the assumption that ordinarily chance has no part in the constitution of the art or culture in question, for there could be no question of voluntarily opening up to something to which one is already constitutively exposed.

The enthusiast for the art of chance procedures, convinced that such procedures produce a different kind of art from other kinds of compositional procedure, can easily be imagined as objecting in the following terms. 'Yes, we can concede that there is an element of chance in all works, indeed in all actions of all kinds. But surely what matters is the degree of indetermination to which a work may be subject? Surely strongly designed or intended works are much more likely to exhibit a determinate and predictable form, over which chance has much less chance to exert an effect, than aleatory works, which are much less likely to be predictable?' To take this line would be to abandon the absolute distinction between absolutely determinate and absolutely indeterminate, but to retain that distinction nevertheless in a logic of approximation in which works are more or less, but to all intents and purposes more, determinate, or more or less, but pretty much mostly, indeterminate.

This way of thinking can indemnify a misunderstanding about the operations of chance and probability. Call it the 'aggregative fallacy'. It is nicely illustrated by an argument developed by Stanley Fish in response to a metaphorical scenario projected by Ronald Dworkin as a way of explaining how it is that judges make decisions based on the history of legal precedents. Dworkin asks us to imagine a chain novel being written by a sequence of authors, each of whom reads what has come before and then contributes a chapter of their own. The first writer, says Dworkin, will be free, because he or she will operate in a field of unconstrained choice – the chapter they write can be about anybody, in any setting, and be written from any point of view and in any style that this frolicking *fons et origo* may hit on. As the narrative is handed on and the collective plot thickens, Dworkin reasons, the choices available will diminish, as successive authors have to take more and more account of the chapters that have already accumulated, until finally, for the last writer in the chain, it may seem as though there is only one possible concluding chapter that can be written.

Fish responds, counterintuitively, that in fact there is no difference, or at least none in terms of the degree of their freedom, between the first and the last in the chain. The first novelist will be free to write what they like, but they will be constrained by everything that is involved in the decision to write a novel, in terms of their understanding of what a novel is and can do. Nor will the putative last or penultimate in the chain have any essential advantage or disadvantage compared with those who have come before, or not, at least, in terms of the ratio between their freedom and their constraint.

For the accumulated pages on top of which they will be sitting will not in fact be self-interpreting, but will themselves need to be construed. The last in the chain will need to interpret what has come before, and will always have the option of radically redefining his and therefore his reader's understanding of what the foregoing novel is taken to be. Indeed, one might very well say that, given the kind of thing a novel is, that is, given the fact that sudden swerves of plot direction, or frame-switching and rug-pulling manoeuvres (it was all a dream, the detective did it herself), are so much part of the horizon of expectation of a novel, such radical reinterpretations may actually start to get more likely the longer the novel goes on. This means that the Johnny-come-lately in the chain is precisely as free and precisely as constrained as its prime mover – that is, his freedom and his constraint are locked together: 'He is constrained in that he can only continue in ways that are recognizable novel ways (and the same must be said of the first novelist's act of "beginning"), and he is free in that no amount of textual accumulation can make his choice of one of those ways inescapable.'[12] We can substitute without significant loss the concepts of determination and chance for the terms constraint and freedom. Whether determination grows or diminishes is itself not a given, but will all, always, *depend* – depend upon the conditions of making out to which the work is subject.

On this estimate, or in this way of conceiving what is involved in the act of estimation, it might sometimes be that the strongly intended or determined work, while being in a straightforward sense more defended against chance, is for that reason more at risk from it. The more set in its ways a novel or artwork may seem to be, the higher the possible yield of innovation or surprise for both writer and reader. Thus a text that may seem to have settled for a place in a comfortable and unchallenging minority niche, giving a modest but regular revenue of pleasure to its fans – *Lady Audley's Secret* as an example of Victorian sensation fiction, say – can be reconstrued by a feminist readership as a searching investigation of the politics of the body, with an interpretative profit in proportion to its unexpectedness which, for that reason, begins to diminish immediately.

What is often seen as a desirable dividend of innovation in artworks – largely because of the horizons of interpretation within which the things picked out as artworks tend to operate, in which sudden changes of meaning and value are themselves a premium source of value – may be seen as an undesirable, even catastrophic, cost if one is talking about a bank or an air traffic control system. It is commonly suggested nowadays, for example, that the immune system of somebody brought up under conditions of strictly controlled hygiene may be unable to cope with the unexpected infectious or pathogenic agents they may later encounter. By contrast, the toddler who has consumed their mandatory peck of dirt and has therefore primed their immune system by exposure to bacterial noise may be much better defended against unpredictable contingencies. We may say that the strongly determined work can have the first kind of immunity. Precisely because it seems so strong, it may in fact be weak at certain crucial points, and in proportion to its strength. The strongly or programmatically undetermined work, by contrast, can come to seem almost immune to accident or the unexpected. In this respect, systematically randomized or aleatory works may be a little like the 1980s TV series based on the work of Roald Dahl that was called *Tales of the Unexpected*, in which the only thing that could ever have unsettled the viewer would have been the failure of an episode to furnish the tediously requisite twist or quirk.

There is another respect in which a strongly determined work may be regarded as more exposed to unpredictability than an undetermined one. For a strongly determined work is very likely to conjoin many different kinds of determination, operating with different degrees of force at different points in the work. The characters might be conventional, but the language obscure and highly wrought; the setting might be stable and unvarying, but the plot subject to lurching time-shifts; and so on. The probabilities in such a work are differentially distributed, in something of the way in which, within a given volume of gas said to be at a certain temperature, there is in fact a distribution of different temperatures, of which the perceived temperature is a statistical average. Precisely because it is determined in so many different ways, and to such different degrees, the strongly determined work is riddled with entry points for chance fluctuations to do their work, sometimes prompting local adjustments to restore equilibrium, sometimes propagating uncontrollably through the system. Such a work constitutes a stochastic landscape, full of chasms and outcrops, slopes, potholes and dimples, in which chance fluctuations might get a toehold.

But this local differentiation is much less likely to be there in a programmatically undetermined work, of the kind that might emerge from the procedure recommended by Tzara, for example. Here, the probabilities, and the improbabilities, are spread out much more uniformly. Since everything is designed to be as unpredictable ('unpredictable') as everything else, we can intuit no landscape or profile of probabilities, no faultlines, no ridges or shaded valleys, no map of mattering. There is more unpredictability on average in such a work, but because the unpredictability goes uniformly all the way down, it is much more predictable at individual points. Where the strongly determined work has many entry points for indetermination, the strongly undetermined work only has one entry point for a difference that would make a difference, which is at the level of the initiating intention to make an aleatory work. This may actually be one of the reasons why aleatory works are so routinely accompanied by a justifying framework explaining the precise procedures employed to produce indeterminate outcomes. The purpose is not to guarantee the paradoxically broken integrity of the work, but actually to make available some point of leverage for the work, since, without the possibility of a difference that would make a difference, the information quotient of the work would be immaculately null. As a result, the apparently predictable work may be more at risk from instability; whereas the giddily unguessable work is in fact metastable, given stability, that is, by the very uniformity of its fluctuations.

The system of the completely aleatory work is like the thermodynamic system that is approaching maximum entropy. In thermodynamic terms, as we have seen, entropy is a measure of the amount of energy that is available to do work in a given closed system: the higher the entropy, the less available work. In thermodynamic systems, the capacity to do work is a function of the amount of organized difference in the system – typically, for example, the separation of hot from cold molecules. The more disorder in the system, the less work can be got from it – you can make a heat engine with a volume of hot gas and a volume of cold gas separated from each other, but you cannot make an engine when those two energy states have been shuffled together. Perhaps this is why the Lord God warns the lukewarm believer that he will be spewed out of His mouth (Revelation 3:16). Order here is not entirely subjective or observer-dependent, for it can be given a mathematical description. An ordered system is one which can be reduced to and generated by a formula that is more economical than the system itself; a chaotic system is one of which the description would offer no possibility of such compression, and would have to match the system exactly. Things drift from order to disorder because, in a given system, the number of ways of being ordered will always be much smaller than the number of ways in which it can be disordered. In moving from order to disorder, therefore, systems move from the less to the more probable, and maximum entropy equates both to maximum disorder and maximum probability. This may at first seem curious, given our tendency to think that disorder ought to be characterized by improbability. The traditional example of a pack of cards can help us over this difficulty. There is only one way in which a pack of cards can be ordered such that the four suits are grouped together and the cards run from ace through to king within each suit. The number of ways in which the cards can fail to achieve this state (52!–1) is huge by contrast and therefore much more likely to occur. So the reason a pack of cards thrown up in the air never seems to come down neatly arranged in suits and numerical values is that there is only one way this can happen, whereas there are billions of ways in which it can fail to happen. This helps to explain why highly disordered states also tend to exhibit what looks like equilibrium; the most likely state for a pack of cards (or, we might just as well say, a bag of letters) that is subject to a series of shufflings is one in which the unpredictability is, so to speak, evenly distributed through the pack.

This can also explain why so many aleatory works are often, frankly, such a chore, since they offer so few genuine surprises, or, better perhaps, their surprises are so reliably and routinely ground out. This might seem to contradict Mallarmé, who declared that 'Un coup de dés n'abolira jamais l'hasard' – 'a throw of the dice will never abolish chance'. After the first roll of the dice, the one that decides that the rest of the work will be generated by rolls of the dice, the scope for chance will be much reduced, precisely because the map of mattering will be so smooth and flat. The maximally randomizing act is like the supreme action-ending act Cleopatra contemplates:

> 'Tis paltry to be Caesar;
> Not being Fortune, he's but Fortune's knave,
> A minister of her will. And it is great
> To do that thing that ends all other deeds,
> Which shackles accidents, and bolts up change.[13]

And yet there is perhaps another sense in which Mallarmé is in fact still right. For the very fact that the pleasures of aleatory works tend to be so insipidly unvarying is not a feature of the works themselves, nor a matter simply of the quotient of unpredictability they contain. It is also a matter of the way in which they function within particular fields of reception, and of how they work out in the different kinds of field in which they are put to work. That is, it is an exposure to relative unpredictability. We are not dealing with closed systems here, in other words, but chained or interlocking systems, in which one system of probabilities is subjected to the force of another.

I'd like to show this by considering what can happen in practice to such claims to have initiated or increased the amount of play in a system. One of the commonest ways in which chance is reified is as a force of liberation or at least of loosening, which can be employed to create new possibilities in a world thought to be otherwise cabined, cribbed and confined by the iron cage of determination and predictability. The dream of such a determinate world and the idea of the liquefying or animating force of 'pure' chance dance cheek to cheek. A good example is furnished by an essay by Natasha Lushetich which describes some of the events that took place in the Fluxus exhibition mounted by Tate Modern in May 2008. Fluxus names a group of artists working in the 1960s and '70s whose work was characterized by the devising of various kinds of event and performative procedure. Lushetich writes in particular about the *Flux Olympiad*, a series of hacked, hampered and otherwise tampered-with games and sports devised by Larry Miller. One of the best of these seems to have been Beci Hendricks's *Stilt Soccer*, which, as its title suggests, requires its players to play soccer while on stilts. The result is a series of improvised methods for trying to retain balance while also pursuing the goals of the game – and, of course, the game will only achieve the desired level of agreeable daftness if the players take it seriously, that is, pretend not to be simply pretending to play football.

This leads Lushetich to the suggestion that the game liberates a 'fundamental undecidability', which, parodying more formalized games and sports, 'restores playfulness to sport and subverts its objectification'.[14] In this, it is said to be representative of a number of such aleatory procedures which dissolve the 'structurality of structure', thus providing 'a nonhegemonic socio-aesthetic practice'.[15] Even allowing for the unhelpful smearing of senses in the term 'non-hegemonic', which could mean either 'non-mainstream' or 'non-authoritarian', though only the first of these is really accurate, this judgement seems unexceptionable. Plainly *Stilt Soccer* is, at least in some respects, a much looser, much less serious kind of proceeding than actual soccer. However, I cannot make much sense of the claim that any kind of 'fundamental undecidability' is involved in this proceeding. First of all, it is governed by rules, just as ordinary soccer is. Indeed it is governed by exactly the same rules that govern soccer played in contact with the ground, albeit combined with another rule, the one requiring the players to walk and run on stilts, that makes all the other rules harder than ever to follow.

In fact, the intriguing thing about *Stilt Soccer* is that it is a perfectly plausible and possibly in time rather a good game, as well as being a witty send-up of one. If the results are unlike soccer as usually played, one has only to observe children who have only just been introduced to the arbitrary restriction of not being able to use any part of their bodies other than their feet to recognize that the way in which *Stilt Soccer* interferes with soccer is a pretty exact recapitulation of the way in which soccer itself interferes with the ordinary ways of carrying and projecting a ball. In short, it imposes a restriction that changes the field of probabilities.

But here is what seems to me to be the salient point. That field of probabilities (the differentially distributed likelihood of being able to control the ball with hands, elbows, feet and head) will itself always operate within other fields of probabilities, which determine (but only partially) the ways in which the activity of soccer will be understood to work. These are often spoken of as defining contexts, but I think they are much better thought of as fields of probability, that may strongly predispose certain ways of understanding as opposed to others, but do not absolutely determine them. We may not need, or even be able, to attach precise numerical values to these probabilities, but we cannot think of them with any kind of finesse without drawing on the mathematics of probability that exercises such a strong interpretative pull on us. A determining context is not one that rules out chance but rather one in which there appears to be a very strong chance that unpredictable things will not occur.

Seen in this way, *Stilt Soccer* could only 'restore playfulness to sport' if it were itself taken to be a sport, or a way of playing it.[16] But what are the chances of this? How many people look to the Tate Modern website for details of soccer fixtures? This is the reason that the event is not a game of stilt soccer, but rather the instantiation of a prankish art-proceeding called *Stilt Soccer*. Typography is here a reliable indicator of typology; one does not go to the Emirates Stadium to see a work with the title *Soccer Match between Arsenal and Chelsea* – except, perhaps, implicitly in those games that are tellingly called 'exhibition matches'. If it caught on, Stilt Soccer would have some chance (though even then a smallish one, I'd say) of restoring playfulness to sports. But *Stilt Soccer* is, on my estimate, vanishingly unlikely to have any such effect, since in general multi-billion-pound sports industries are not much affected by developments in the fields of art practice and aesthetic theory. There is obviously a certain kind of playfulness in *Stilt Soccer*, but that playfulness is quite strongly fenced in by where and how it occurred (in a place where art happens). How far it can restore playfulness to anything will depend upon how that playfulness is itself put into play, or, as we say, played out, in different fields of expectation or probability.

This leaves – though one had as well say 'preserves' – the possibility of writing about chance not as topic (novels about gamblers), or as strategy (the work of art employing chance procedures), but as universal yet universally variable condition, or, indeed, the condition of universal variability. Chance would then be regarded, not as something internal to the artwork, that is, part of its theme or content, or something outside it to which it is thought to be dangerously or thrillingly exposed, but as threaded through the very working of the work itself, as it is put into play. Chance is not on the other side from determination, it is the very process whereby determination and chance are distributed. Determination and chance are not to be put into separate piles and simply totted up, since the force of determination that a work will seem to exercise or exhibit will itself be a function of chance.

So we might do well to avoid the bipolar mood swings of absolute choice on the one hand and absolute chance on the other, and learn to inhabit what Gary Saul Morson has followed Aristotle in calling 'causality for the most part'; as Morson tellingly observes, 'Books may be called *Chance and Necessity*, as Jacques Monod's famous one is, but I have never seen one called *Chance, Necessity, and For-the-Most-Part Causality*.'[17] Morson is right to condemn the Leibnizian or Laplacian determinism that governs ways of seeing society and history. The name of Laplace is anyway unfairly latched to determinism, given his great importance in the history of probability theory. But, even if Laplace is absolved of the blame for determinist thinking, Morson is right to call this way of thinking 'crypto-theological'.[18] Oddly enough, the humanities and social sciences, though recoiling from the forms of quantitative thinking characteristic of the exact sciences, and proclaiming their difference from them in their embrace of the undetermined and indeterminably particular, in fact assume and inhabit a world of absolutes that has seemed laughably unlikely since Laplace. It is often suggested that human affairs are not to be understood with the reductive models developed for understanding processes in the physical world, because those affairs involve a multi-parameter calculus that is too large and intricate to undertake with any hope of success. Yet this claim cohabits in the analysis of social forms and processes comfortably with the explicit or implicit dependence on models that, while they are nearly always derived from the analysis of the natural world, are in fact far cruder and more approximate than any of the models that have been developed to deal with physical processes over the last couple of centuries. These metaphor-models are, for example, geological ('strains', 'faultlines', 'eruptions'), or hydraulic ('currents of influence'), or meteorological ('prevailing climates of opinion') or crystallographic (complex symmetrical structures of every kind). But such models would scarcely suffice to describe and predict the bobbings of a rubber duck in the bath, let alone the movements and tendencies of human affairs. If these affairs are really as complex as we know they must be, why cling to such clunky, clanking machineries of mind to model them?

This dependence on the most reductive kinds of models permits us to permit ourselves to confuse precision and clarity. As we saw in Chapter Two, precision in fact requires indeterminacy; it makes fuzziness unavoidable. Absolute clarity, by contrast, depends upon approximation. Strangely, then, it is the inexact sciences that depend on absolutes, and the exact sciences that have long recognized the need to operate without them. The less exact you are, the more absolute you may allow yourself to be; the more exact you are, the less absolute you can afford to be.

THE WEATHER OF HISTORY

I have begun by speaking about the analysis of individual works, but this is in order to broach a way of thinking that would apply on a much larger scale, to the forms of organization we call cultures, particularly as they may be conceived historically. If the oscillations of a planet's magnetic field, the prosody of a dripping tap, the formation of a snowflake, are to be understood as stochastic processes, why should we expect human history to line up with the crudely deterministic models we deploy upon it?

Cultures are sometimes represented as organized sets of principles, articles of faith which can be plainly articulated, along with the systems of behaviour to which they give rise – 'Protestant cultures value individuality'. But such beliefs and behaviours are never in fact uniformly adhered to, or even in fact universally mandated, in a particular culture, though there may be strong pressures towards them. Cultures are best thought of as climates – climates not only of opinion but of feeling, belief and action. To be French or female or fin-de-siècle is to inhabit, and to contribute to, such a climate. But that climate is not a given, but rather a set of potentials or probabilities, values towards which things in that setting will tend. This means that cultures, like climates, are unlikely to have anything uniquely and finally distinctive about them. It may be that, as has often been said, a text like *Hamlet* marks the beginning of a particular style of intense self-consciousness that had not previously been part of the way in which individual human beings saw and thought of themselves, but that from the beginning of the seventeenth century onwards would begin to be more common. But it is implausible and unhelpful to imagine that nobody could ever have felt anything like Hamlet's squirming reflexivity before that, or that 'the subject' was born at that moment. This is not just because the transitions from one era, or prevailing structure of feeling, to another are slow and irregular. It is also because there is no absolute reason why a Hamlet-like self-relation might not have arisen in any other period whatsoever. It may have been quite unlikely, but it could have occurred; indeed, given the numbers of persons and occasions involved, it must have, and often. We might think of the possibility of such events as we would think of the possibility of unlikely climatic events, say, snow in Sydney. The lowest temperature ever recorded in Sydney was 2.1°C, and the last recorded snowfall was in 1836 (though there are sticklers who insist that this can have been no more than airborne slush). But this is not enough to justify the assertion that it never snows in Sydney, only that it is very, very rare for it to do so, and pretty unlikely that it will do so in the next, say, hundred years. Cultures are like spread bets, complex probability profiles.

Indeed, just as weather systems are not simply independently occurring events, but tend to feed back on and amplify themselves, a high-pressure system may prove to be very stable, making it temporarily much more likely that the weather tomorrow will be the same as today than it usually is – so cultures will not only make certain events more likely, but will iterate those events. Indeed, what we mean by 'a culture' may best be thought of, not as a field of likelihood of certain events happening – the appearance of Hamlet – but as fields of selective attention that are much more likely to pick out certain kinds of event as significant than others.

Still, one might say, as with the weather, that these fields of probability, though they are always in operation, are necessarily always prospective. Nobody bets on a race that has already been run. Time runs from the soft to the hard, from the indefinite to the definite, the virtual into the actual. Time continually condenses probability into positivity, possibles into givens, fractions into cardinal integers. The process of time means that things that did not have to happen keep happening, and however unlikely they were, they thereafter will always have happened, though without having had to. This process itself has something like the force of a necessity; it has to happen that things happen that do not have to. Non-necessity is necessary.

This may appear to make it mean that probability considerations actually have no place in the retrospective constructions that we call history. Of course, it may certainly be that what we find of interest in a given historical field is strongly influenced by the probability gradient of our attention, as we selectively pick out and amplify certain kinds of feature in the field of givens. But it is hard not to believe that, even though we are exceedingly unlikely ever to be able to access and know it, there was a 'fact of the matter' about everything that has already happened. The relations we establish with the past may be variable, but that with which we seek to establish a relation is surely not. The race has already been run, and, no matter how we reinterpret the outcome, it can never be rerun and the outcome decided differently.

History gives every appearance therefore of going from the soft to the hard. We think of facts – dates, data – as fixed and finite, and relations as potential, infinite. There is no limit to what may be made of the Battle of Trafalgar, and no definitive way of estimating the odds on any particular way of making sense of it, since these will depend on preoccupations, and significance-amplifying conditions that have yet to arise.

This suggests that facts can never be natural events. What's done is done, and can never be undone. But the question of *what* it is that has in fact been done is not one that can easily be done with. So, in a certain sense, it is facts that are soft, not relations, or rather it is relations that make occurrences into hard facts. Nothing that has simply happened, that is to say happened without having entered into some relation of significance, has really happened yet, for there is as yet nothing for or in terms of which it could be a fact. Things that have happened only once have not yet happened at all. Only things that have happened twice – once in the mode of occurrence, and then again in the mode of recurrence, have happened and then been seen to have happened – have happened in the first place. This is not to deny that the events of history have taken place, but it is to say that these events are potentially infinite, depending as they do on the future, and therefore without significance.

This means that there can also be no such things as the unique and wholly unpredictable 'events' that contemporary philosophers such as Alain Badiou evoke. However rare and precious these atoms of incident may be, they can only be events insofar as they have already been put into play, retrospectively constituted as events, in the fields of probability to which they are subjected. Like everything else, events must take their chances. We think we are on the determinist side of events that have moved from the virtual to the actual, but we are always in fact between the virtual and the actual, the determined and the undetermined. We are like the player who, just having gone down 3–2, suggests: 'best of seven?' Our situation, and the situation outlined by Stanley Fish, with respect to the chain-novel of history, resembles the 'problem of points' investigated by Pascal and Fermat, in the correspondence which inaugurated modern probability calculus – that is, the problem of assessing the likely outcome of a game which one is in the middle of playing.[19] What Pascal and Fermat could not take into account was that their way of estimating the just outcome of the game might in fact be part of the game.

So is it ever game over for the past? No, because the relations which we convince ourselves constituted a text were only ever a partial actualization of its possibility, which it may be left for us or others, depending perhaps largely on our calculation of the likely profit or utility, to attempt to complete. It will remain true for ever and a day that on Thursday, 16 June 1904, Throwaway won the Ascot Gold Cup by a length from the 5–4 on favourite Zinfandel, at odds of 20–1. No matter how many times the race might be replayed, Zinfandel will never make up that length to alter the outcome of the race. But the significance of that fact currently seems to have a better-than-average chance of continuing to radiate and unfold. Why? Because, lucky as Throwaway and his backers might have thought themselves, they got luckier than every other horse, and every other race that day, because this race was picked out by James Joyce for special attention in his novel *Ulysses*, and, presumably, by using something of the process that punters themselves might have used, namely, scanning the papers and looking for some circumstance in the name of the horse that seemed to pick it out from the rest and promote it to attention – something, in other words, that reduced its randomness and increased its redundancy.

Throwaway survives because Joyce won his bet, meaning that it was able to become multiple, entering into a relation that will ensure continued replication (though only and exactly for as long as it does). This is to say, it continues in being by entering into a field of probabilities, which could not reasonably have been thought to be a likely part of its original field of opportunity. Things survive, seeming to exhibit in the very fact of their survival a will to persistence, because they are repeatedly selected, because there is a high chance of their seeming significant.

This is not to say that the past is endlessly revisable or retractable. This view does not require anything like an alternative universes theory, which would allow us somehow to throw the dice again and get a different outcome. There are facts, but those facts can be made meaningful in many different ways. An historical fact is like a move in a game that is still in process. Because the game is not over until the fat lady sings, and there is rarely any sign of the approach of the taxi bearing this consummating personage, the significance of any particular throw is not yet completely settled. A statement about the past is always hazardous, to the degree that it is a prediction about its future. A fact is only an event with a high probability, so far, of being replicated without modification, in changing circumstances.

I have been trying to show in this chapter that it is unhelpful to think of chance as outside or beyond determination. I have also suggested that this makes for a complex interlacing of before and after, anticipation and retrospection, in history, which articulates time, in both its senses, dividing and connecting it. Number does not govern or completely account for this process, but provides a subtle and sensitive form of access to it. The kinds of quantitative method which are beginning to become available, as digitization and new methods of data analysis provide us both with more forms of evidence and ways of investigating it, can only help us to develop more fine-grained and verifiable accounts of fields of meaning and the conflict and circulation of concepts as a result.

But chance is different from all the other kinds of otherness, or has now historically (that is, by chance) become so. For centuries, chance has represented the incalculable as such, absolute, unknowable and intractable. This left the sphere of causality, determination and the calculation of consequences intact. If simply no account can be taken of chance, then it can be set aside as the simply or absolutely unknowable, meaning that, reduced as its sphere may be, one can at least count upon what one can under no circumstances calculate.

What happens from the seventeenth century onwards is the entry of chance into calculability itself. The advantage of this is clear, and the vast importance of statistical reasoning in contemporary life makes it obvious how impossible it would be to do without probability calculus. Many calculations actually depend upon randomness, hence the strange and paradoxical quest for reliable ways of generating genuinely random numbers, in other words for a determinate indetermination.[20] The cost is the surrender of the possibility of exactness, even as an ultimate horizon. Henceforth one works not against error and inexactness, but with and within them. It is not the incalculability of chance that is the problem, but the fact that it is no longer possible to regard chance as wholly incalculable and remain honest.

8

KEEPING THE BEAT

In *The Art of English Poesie*, George Puttenham introduces a contrast between strict metre and a more flowing kind of rhythm, the second of which he characterizes with the term *numerosity*:

> There is an accomptable number which we call *arithmeticall* (*arithmos*) as one, two, three. There is also a musicall or audible number, fashioned by stirring of tunes & their sundry times in the vtterance of our words, as when the voice goeth high or low, or sharpe or flat, or swift or slow: & this is called *rithmos* or numerositie, that is to say a certaine flowing vtteraunce by slipper words and sillables, such as the toung easily utters, and the eare with pleasure receiueth, and which flowing of words with much volubility smoothly proceeding from the mouth is in some sort harmonicall and breedeth to th'eare a great compassion.[1]

Puttenham stresses that numerosity is a matter of gliding rather than hopping – as we might now sometimes say, of continuous or analogue movement characterized by ease, fluidity, smoothness and volubility, rather than discontinuous or digital movements. But *numerosity* makes it clear that this movement is not unnumerical, but rather 'numerous' – numerous without being numberable. It is, he says, *rithmos*, which is not exactly arithmetical, but, since inexactness, a certain kind of uncertainty, is exactly what seems to be in play here, it is not exactly *not* arithmetical either. Puttenham assumes that *arithmos* means something like 'without flow', though there is in fact no direct relationship between the words *arithmetic* and *rhythm*. *Rhythm* derives from ῥεῖν, to flow, that is audible in words like *rheostat*, *rheum* and *diarrhoea*. But *arithmetic* is not a privative form of *rithmos*, as Puttenham wants to think, but derives more directly from ἀριθμός, *arithmos*, a number, itself deriving from the verb αἴρω, *airo*, meaning to raise, lift or take up, so presumably involving an idea of the reckoning up of number. There cannot be a pure rhythm, pure flow, since there can only be flow across what breaks or retards it. Amps equal volts divided by ohms. Without ohms, no amps: zero resistance, no current. To go with the flow completely is to be static; the foot must step into the river to feel the force of its current.

The word *numerosity* appears at irregular intervals in discussions of prosody and versification over the next few centuries, often to characterize the move from the unrhymed and accented verse of Greek and Latin poets to the more formal metres and rhymed verse of English poetry. In these discussions, numerosity usually means subtly differentiated but harmonious flow. Oliver Goldsmith writes that 'cadence comprehends that poetical style which animates every line, that propriety which gives strength and expression, that numerosity which renders the verse smooth, flowing, and harmonious.'[2] Edward Wadham deploys the word to articulate the traditional preference for the subtlety of Greek poetic rhythm in comparison with the plonking regularity of rhyme and metre in English verse. Wadham argues that 'Music is as the vowel, rendered infinitely impressionable by being entirely dissevered from the consonant,' such that, in a melodious passage like the beginning of the *Odyssey*, 'the instances of half-rhyme, alliteration, and every variety of approach to repetition of foregone sounds, are absolutely too multitudinous to indicate; the whole verse is alive with their playing and combining, like a sunset sky with irradiate tints.'[3] Rhyme is a blocky contraction of rhythm, a mechanical squashing together of what the melodious rhythm of numerosity generously lengthens out:

> Rhyme is too intense for melody; it is the caricature of it, nothing more. Suppose it, say, the aggregation of melody into one spot — but is melody a thing that can be aggregated? Melody is rather numerosity, a blending murmur,

> than one full concordance. Melody is as effectually silenced by rhyme as the tones of a flute under the beating of a drum.[4]

Numerousness could occasionally substitute for *numerosity*, for example in Dryden's praise of Horace in 1685: '*That which will distinguish his Style from all other Poets, is the Elegance of his Words, and the numerousness of his Verse; there is nothing so delicately turn'd in all the Roman Language*,' though this is a rather unusual use of the term.[5] In a 1762 essay on Greek and Latin prosody, John Foster makes an explicit distinction between 'numerosity' and 'quantity', in arguing that appreciating the harmony of Latin verse requires that one pay attention to accent as well as beat, or 'tones' as well as 'times': 'Those, therefore, who, in considering the numerosity of writings, attend to quantity alone, regard only the inferior part of the subject before them.'[6] A writer on education in 1785 observed that, in proverbs, 'the numerosity of the sentence pleased the ear and the vivacity of the image dazzled the fancy.'[7]

Not everybody thought of numerosity as subtlety. Thomas De Laune heard in the rhythms of Greek and Latin verse an '*idle or delicate itch of Words, that external sweetness or allurement, that numerosity of sounds, or those pleasing trifles*' that was not evident in the '*Grave and Masculine Eloquence*' of the sacred scriptures.[8] And sometimes the word 'numerosity' was used in the opposite sense to that distinguished by Puttenham. Thomas Sprat uses it to mean the close adherence to regular metre in his discussion of Cowley:

> He understood exceeding well all the variety and power of Poetical Numbers; and practis'd all sorts with great happiness. If his Verses in some places seem not as soft and flowing as some would have them, it was his choice not his fault. He knew that in diverting mens minds, there should be the same variety observ'd as in the prospects of their Eyes: where a Rock, a Precipice, or a rising Wave, is often more delightful than a smooth, even ground, or a calm Sea. Where the matter required it, he was as gentle as any man. But where higher Virtues were chiefly to be regarded, an exact numerosity was not then his main care.[9]

Thomas Morell borrows Sprat's phrase to defend Chaucer's versification in his edition of *The Canterbury Tales* in 1737, remarking that 'an exact Numerosity . . . was not Chaucer's main care . . . His Numbers, however, are by no Means so rough and inharmonious as some People imagine.'[10]

Despite these variations, numerosity seems to be used to name something like the principle I have been calling quantality in this book. It seems to evoke a counting that is not, in Puttenham's term, 'accomptable', that cannot quite keep count of, nor yet quite account for, itself. Nowadays, 'numerosity' has acquired the meaning among psychologists and animal ethologists of the 'number sense' as it may be displayed in humans and other species.[11] I have not yet found evidence of this usage much before the 1990s (and the OED does not yet even register it).

Number is a kind of irritant within the history of thinking about rhythm and metre in literary studies. There is a tension in this history between the effort to account with complete and objective precision for prosodic structures and effects, thereby reducing literary effects wholly to a matter of number, and accounts that focus on the complexity of the interinvolvement of objectivity and subjectivity in the apprehension of rhythm. Simon Jarvis, who has called urgently and sustainedly for prosody to be taken seriously as part of the distinctive kind of cognition that poetry offers, has argued that the dream of making prosody fully accountable, to itself or to anything else, is a sort of myth or mania: 'As well crack quantum mechanics upon the Roman Rite as set linguistics to a total calculus of metrical types . . . prosody cannot be grounded on the model of the measurement of an object.'[12] Jarvis believes that one cannot simply and serially correct the mistakes of past theorists of versification with a more precise scientific method because 'the "mistake" is this idea which the scientistic prosodist has that his or her method is a fully demythologised one.'[13]

But, in his opposition to the dream of full scientific explicitness, the verbal equivalent, perhaps, of accountability, Jarvis perhaps attempts to keep number at bay in an absolute way that presents difficulties. This is precisely because of the claims he wishes to push through for the role of prosody in thinking and in the formation of the subject. Jarvis quotes Henri Meschonnic to the effect that 'There can be no theory of rhythm without a theory of the subject, and no theory of the subject without a theory of rhythm,'[14] and finds warrant for this in Hegel's description of rhyme as a figure for the coming back of the subject to itself in time:

> The subject is a self-exteriorization and a return, a recollection after an excursion, for which language furnishes the most eminent model, but which is also seen, for example, in the structure of human labour. Only this excursion and return can convert the merely indifferent flow of time into the shaped and understood duration which makes subjectivity intelligible.[15]

Jarvis concludes that 'it can in a certain sense be said that the subject rhymes, for Hegel.'[16] This is a brilliant intuition. Jean-Luc Nancy has argued for something like the same structure of resonance in the formation of the subject through the reflexivity of rhythm:

> We should linger here for a long while on rhythm: it is nothing other than the stroke of time, the vibration of time itself in the stroke of a present that prevents it by separating it from itself, freeing it from its simple stanza to make it into scansion (rise, raising of the foot that beats) and cadence (fall, passage into the pause). Thus, rhythm separates the succession of the linearity of the sequence or length of time: it bends time to give it to time itself, and it is in this way that it folds and unfolds a 'self.'[17]

Nancy offers us a resonant subject in place of a subject conceived as a line of sight or point of view, which is

> an intensive spacing of a rebound that does not end in any return to self without immediately relaunching, as an echo, a call to that same self. While the subject of the target is always already given, posed in itself to its point of view, the subject of listening is always still yet to come, spaced, traversed, and called by itself, sounded by itself.[18]

This is to make the self a particularly musical kind of counting out or self-enumeration. The I becomes itself, is able to be at one with itself, by dint of a self-division, by going out from and coming back to itself. In this way, it follows a numerical logic we have met with repeatedly in this book; it becomes one by becoming two, since it is only by being the kind of thing that can be counted as one, that is that could be counted twice, that the one becomes knowable to itself. I have said it before: there must be at least two before there can be one.

A rhythmic self is a self that, at least in part, comes to itself through this kind of counting, or self-enumeration. I think that closer and more candid inspection will reveal that sound, self and the sense of number are tightly articulated with each other. A much-repeated remark of G. W. Leibniz may get us in the way of thinking about this. In a letter of 17 April 1712 to Christian Goldbach, Leibniz wrote that 'Musica est exercitium arithmeticae occultum nescientis se numerare animi.'[19] Oliver Sacks offers a pleasingly rangy translation of this in his book *The Man Who Mistook His Wife for a Hat*: 'The pleasure we obtain from music comes from counting, but counting unconsciously. Music is nothing but unconscious arithmetic.'[20] This chimes kindly on the ear, though the 'nothing but' is an extra quantity and there is really nothing in what Leibniz says about pleasure, even if one might reasonably assume that the *exercitium arithmeticae*, the arithmetical exertion of the mind, might indeed be a source of enjoyment. Rather more precise is the translation offered by E.F.J. Payne in his English version of Book III of Schopenhauer's *The World as Will and Representation*, in which he quotes Leibniz: '[Music is] an unconscious exercise in arithmetic in which the mind does not know it is counting.'[21] But it is still not unimpeachably precise. *Occultum*, 'hidden', may imply unconsciousness, but is in fact something different from *nescientis*. It may certainly be that the exercise is hidden simply because the mind is unaware that it is exerting itself in this way, but it might well be hidden in some other way.

Leibniz repeats these sentiments in his 'Principles of Nature and Grace' of 1714:

> What is more, even the pleasures of sense are reducible to intellectual pleasures, known confusedly. Music charms us,

> although its beauty consists only in the agreement of numbers and in the counting, which we do not perceive but which the soul nevertheless continues to carry out, of the beats or vibrations of sounding bodies which coincide at certain intervals. The pleasures which the eye finds in proportions are of the same nature, and those caused by other senses amount to something similar, although we may not be able to explain them so distinctly.[22]

If Leibniz is right and all music involves some kind of counting, it would be worth our while to wonder if there were reciprocally some music in all counting. It would seem at least to be the case that there is always something auditory in counting, precisely because counting is what enables us to stretch or exert ourselves beyond what can be grasped by the native visual numerosity that enables us to grasp and name small numbers. It is not possible to count items of any kind without engaging in some kind of recitation, some sort of counting out, where the 'out' is the extension or exteriority of time rather than of space. And, extension being related as etymological cousin to Greek *tenein*, to stretch, to tone and tune, this kind of temporal extension may be thought of as a form of intonation. Leibniz seems to be pointing to a kind of numerosity, a toning or entraining of the mind, that numbers enact without explicit counting.

As we saw in Chapter Two, Sacks in fact sees this kind of inner counting as part of what knits brain to body. Someone who has been deprived of the use of a limb may find that it drops out of their body map, and must be actively reincorporated. In fact, 'body map' is perhaps a bit of an approximation here, for it is not so much a picture of the limb that is required, especially if it is a leg or foot that is usually experienced in motion, so much as some kind of bodily melody that must be remembered for the limb itself to become oneself again a member of the orchestrated body. Melody seems better than map, because what one must remember in walking is not something one has to grasp, or keep in mind, like a picture or diagram, but something one has to know how to do without knowing quite how one is doing it. Walking is like saying your times tables, playing a scale or reciting a poem. Sacks describes how, following the climbing accident discussed above in Chapter Two and the forced immobilization of a leg for a fortnight, he had forgotten the rhythm of walking. The Mendelssohn Violin Concerto in E Minor, which he had been playing repeatedly in hospital, came to his aid:

> Suddenly, as I was standing, the concerto started to play itself with intense vividness in my mind. In this moment, the natural rhythm and melody of walking came back to me, and along with this, the feeling of my leg as alive, as part of me once again. I suddenly 'remembered' how to walk.[23]

Not the least striking thing about Sacks's account is the fact that the leg is here brought to life, turned from a dead, merely mechanical appendage into something incorporated, by means of something itself inanimate or mechanical, the music which, by dint of the fact that Sacks had been playing it over and over on the only cassette tape he had with him, started 'to play itself' in or through him. A little later, Sacks describes a woman with a paralysed leg following a hip fracture, whose leg could not move at all, except once when 'it had kept time at a Christmas concert, "by itself," when an Irish jig was being played'.[24] The rhythm of music seems to induce motion, purely and immediately through motion itself. Music seems to be an excellent way of structuring and storing information sequentially. The songs and melodies that most of us know, as well as information embodied in rhymes, code for a kind of knowledge that we cannot hold in our minds or access all at once, but must allow to play out diachronically in the mode of counting through: 'Thirty days hath September . . .'.

But this points to an interesting feature of counting, which may complicate the contrast that Puttenham offers between rhythm and arithmetic and the distinction offered by Leibniz between conscious and unconscious counting. For counting cannot perhaps be said to be something that we can ever do entirely consciously (even supposing we could attach some coherent meaning to the idea of being entirely conscious). We are able to count only if we have learned already how to count, that is, if we already have names for the numerical values that occur in ordinal sequence, and know their sequence. I do not have to think much about counting, any more than I have to think about reciting the alphabet, because they are both autonomized routines. As I say 'five', 'six' seems already to be welling up in it, and I cannot in any sensible way be said to 'decide' the matter of what number is to follow. The transition seems to happen in a sort of cinch or pocket of time, and these blinks of the mind are what propel me along the automatic sequence of counting. So, if it is true that we count things out in order to try to be more precise or explicit about how many items they may consist of, it is also true that there is something that is implicit, puckered or precoded in the action of counting. I can only count because I do not have to account fully for everything in the process of counting and so cannot be said ever to be fully conscious of it, or fully and equally at hand at every moment of it. Oliver Sacks, drawing on the work of many others, points to the likelihood that humans 'are the only primates with such a tight coupling of motor and auditory systems in the brain – apes do not dance, and though they sometimes drum, they do not anticipate a beat and synchronize to it in the same way that humans do'.[25] This may imply that there is indeed a kind of 'hidden exercise' involved in music, but that the exercise or putting of oneself into play is not one of which one could be completely the pilot. Exertion is from *ex* + *serere*, an unbinding, where *serere* means both to join, bind or intertwine, whence *series*, and also to sow, strew or spread, whence *serum*, *semen* and *dissemination*. When I count as I listen to music, whether consciously or unconsciously, I make the music conform to some external framework. But it really is beginning to seem as if there is a kind of music involved in counting itself, which partially swallows me up. Counting is always, in some measure, dance, or trance. The relation between dance and music, each one a kind of register or index of the other, is itself one of reciprocal measure, as when the Duke urges the couples joined at the end of *As You Like It* 'With measure heap'd in joy, to th'measures fall.'[26] There is always a kind of falling in (cadence) incident to musical counting.

This becomes particularly apparent in the practice of 'counting in', whether it be sounded out, or tacit, in the beats of the conductor's baton. In this process, the count is set going, as an autonomous mechanism. As the music begins, it blends into the introducing count, which blends into it. It is as though, taking up the count, the music begins to count for itself, immanently.

Counting can sometimes seem to become fully autonomous, giving itself its own law. In his account of his delusions during a period of madness and incarceration, John Perceval describes a counting routine that got out of hand in this way:

> Weary at length, and unable to comprehend these commands, I sought for sleep, and recollecting what my mother had formerly told me of my father, that he used when he found himself unable to obtain rest, to keep continually counting to himself, I tried the same. But then the power of thinking numbers for myself was taken from me, and my mind or life lay in my body, like a being in a house unable to do anything but listen to the sound of others talking around him, and voices like the voices of females or fairies – very beautiful – very small, and with a rapidity I cannot describe, began counting in me, and entirely without my control. First, one voice came and counted one, two, three, four, up to ten or twenty – then a second voice took up the word twenty, and kept repeating twenty – twenty – twenty – whilst another after each twenty called one – two – three – four, and so on till they came to thirty – then another voice took up the word thirty, and continued crying thirty – thirty – thirty, whilst a voice called out after each thirty – one – two – three – four, and so on till they came to forty, and thus the voices within me proceeded, dividing the labour between them, and so quickly, that I could not possibly pronounce the numbers.[27]

This sounds like a kind of literalization of an experience of one's thoughts racing away which is actually quite common, especially in insomnia. But it also tilts us towards the little, lulling slumber that lurks in every effort at 'thinking numbers for myself', in Perceval's intriguing phrase. It is not clear whether there is suffering or joy in this delusion. The counting eventually gets beyond Perceval's powers to follow it, so he is himself unable to keep count of the counting that is going on inside him. Yet, perhaps partly because of this fact, he pronounces the sound 'very beautiful' and, if he thinks of himself being subjected to or becoming assimilated to some kind of machinery, that machinery is female.

Richard Feynman describes a rather more willed and systematic series of experiments in internal counting. Prompted by reading an article about variations in the time sense induced by the experience of fever, Feynman established through repeated practice a kind of internal metronome that meant he was able reliably to count up to sixty, taking very close to 48 seconds every time. Having trained up this internal counting module, Feynman started to experiment with things that might disrupt it. He proved to be able to read out loud perfectly competently while maintaining his internal count, though he found it difficult to perform other counting operations, such as running up and down stairs or counting items in his laundry, at the same time:

> when I put out the laundry, I had to fill out a form saying how many shirts I had, how many pants, and so on. I found I could write down '3' in front of 'pants' or '4' in front of 'shirts,' but I couldn't count my socks. There were too many of them: I'm already using my 'counting machine' – 36, 37, 38 – and here are all these socks in front of me – 39, 40, . . . How do I count the socks?[28]

Feynman does not mention listening to or playing music while counting, but the capacity of percussionists to maintain different rhythms simultaneously suggests that this might not have made for insuperable difficulty. Listening to music seems to involve what Puttenham calls a 'compassion', in which it is not quite clear how action and passivity are distributed. When I count, I seem consciously to be regulating something that would otherwise go beyond my control. When my mother, who never drove or owned a car, would stand helpless and infuriated at the kerb watching the cars sweep by, she would channel her rage into counting them: 'one, two, three, four, *five*, SIX', she would count out, six marking the appalling limit of intolerability. Counting through to six allowed her a triumphant declaration of the unspeakability of being made to wait that long. But at the same time, I give myself over in counting to something that seems to be counting itself out through me. The point about counting seems to be that it is never clear who or what exactly is doing it. Why else would one count in order to go to sleep? The 'exercitium' in Leibniz's formula is conductive, passing from the music to which I lend my ear, to me, and then back out again to the music which seems to me to be inciting my exertion.

All this is a question of rhythm, a word which, like the word *series*, conjoins measure and flow. There can be no rhythm without flow, without the movement or exertion that carries one across from one beat to another, yet equally there can be no flow without the beats or divisions between which the flow occurs. There must be matter for there to be metre; there must be the hard for there to be soft.

SPANKING AND POETRY

For this reason, there is also a certain measure of cruelty, the cruelty of measure itself, in rhythm. The figure of the conductor has been a late arrival in music, and, where there was a leader of an orchestra, he might very well confine himself simply to beating time. Carroll's Mad Hatter's tea party provides a silly but telling reminder of the agonistics of time-keeping:

> 'I dare say you never even spoke to Time!'
>
> 'Perhaps not,' Alice cautiously replied; 'but I know I have to beat time when I learn music.'
>
> 'Ah! That accounts for it,' said the Hatter. 'He wo'n't stand beating. Now, if you only kept on good terms with him, he'd do almost anything you liked with the clock . . .'[29]

The most well-known story of the violence and fatality that lie latent in the beating of time is that of the death of Jean-Baptiste Lully, who is said to have accidentally struck his toe while beating time with his staff during a performance of his admittedly strapping *Te Deum*. The toe became infected, and, refusing offers to save his life by amputating it, Lully died of gangrene.[30] The history of conducting involves a move from the hard to the soft, from conducting as driving (*conduire* is the French verb for 'to drive a car') to conducting as subtle interchange of energy. The baton no longer strikes some surface audibly: there is no object for it to come up against in the form of an audible impact, or at least not after the imperious rap on the music stand which is the only vestige of that uncouthness. Instead, there is a nervous, quivering but infinitely powerful shaping without touching in air, making out a mobile topology of feeling, a third space between orchestra and conductor in which the music can be figured, leading and led by it. In this space, positions and directions are mingled and transformed, conducting becoming induction, reduction, production, seduction.

Eve Kosofsky Sedgwick has pointed to a similar interchange between the hardness of compulsion and the softness of wanting in her essay 'A Poem is Being Written'. This most earnest and formidably authoritative of literary critics begins her essay with the bold and altogether breathtaking announcement that 'When I was a little child the two most rhythmic things that happened to me were spanking and poetry.'[31] The episodes of family spanking, and the memories that recall them, constitute a 'breath-holding space' (*Tendencies*, 182), in which control and release are held in tension, and the rawly shameful excitement of exhibition coexists with sonorous and tactile rhythm:

> A primal hunger to be seen was certainly not undone in these punitive moments, but only made inseparable from the paralysis of my own rage and the potency and bland denial of my parents' rage; from the tensely not uncontrolled, repressed and repressive (and yet how speaking) rhythm of blows, or beats; from the tableau itself. (*Tendencies*, 182–3)

There is a rhyme, amounting almost to magical identity, between the scene of spanking and the similarly apnoeic poisedness of the lyric poem, a containment which is a space, but also a stretch, marked out by beating or pulsing, in which percussion and pulsation may take place:

> The lyric poem, known to the child as such by its beat and by a principle of severe economy (the exactitude with which

> the frame held the figure) – the lyric poem was both the spanked body, my own body or another one like it for me to watch or punish, and at the same time the very spanking, the rhythmic hand whether hard or subtle of authority itself. (*Tendencies*, 184)

This beating is an alternation of scene and sound, the iambic thud of the rhythm programming a rhyme between the alternation of sound and silence and the alternation between movement and stasis, even extending to the segmenting of the child's stripped body: 'The glamorized, inbreathing theatrical space of the spanking thus contracted to the framing of a single, striped, and sectioned midbody that wanted to move and mustn't' (*Tendencies*, 183).

The essay, which intersperses extracts from poetry read and written at different stages of Sedgwick's life with reflections on spanking and the pornographic imagination (candidly her own), even proposes a link between chastisement and the poetic technique, which she discovered early on, of enjambment.

> I knew *enjambment*, not just for a technical word in the introduction to my rhyming dictionary, but for a physical gesture of the limbs, of the flanks, the ham. I thought then, too – in fact I thought it until I checked my dictionary just today – that a *doorjamb*, for instance, was the thing one wedged in the door to keep it open, a doorstop. From all this I visualized *enjambment* very clearly as not only (what my French dictionary now tells me) the poetic gesture of *straddling* lines *together* syntactically, but also a pushing *apart* of lines. In terms of the beat(ing) of the poem, enjambment was, in this fantasy that shaped my poetic, the thrusting up out of the picture plane in protest by the poem's body of a syntactic thigh or shank that would intercept, would retard the numbered blow: would momentarily wedge apart with sense the hammering iteration of rhythm. (*Tendencies*, 185–6)

Enjambment is a wedging or delaying – the leg raised to intercept the blow – and also a crossing or straddling. Like the two-stroke lub–dub of iambic poetry, or, come to that, of spanking, it both arrests and accelerates, in a stutter of fluidity. Like poetry, ballet is 'a rhythmic, prestigious, exhibitionistic and highly theatricalized way of choosing the compelled and displayed body' (*Tendencies*, 186). And Sedgwick insists that the allure of this spectacle was the compounding in it of pained passivity and impassioned choice, discovering for herself, as many children subject to violence from which they cannot escape may discover, how 'to abstract the body of one's own humiliation; or perhaps most wonderfully, to *identify with* it', so that 'the *compelled* body could be *chosen*' (*Tendencies*, 184). Under these circumstances, or at least in their recollection, it is the suffering child who is 'ravenous for dominion', a dominion (but whose?) to which the writing of poetry gives her (*Tendencies*, 184). And the recollection here, as Sedgwick slyly hints, is in any case the excuse or occasion for fantasy: the intensity with which it is remembered seeming to be in proportion to the unlikelihood of its ever having taken place.

Yet it is disappointing and indeed a little perplexing that 'A Poem is Being Written', which seems to insist so much on the importance of rhythm, is almost totally anacoustic. If there is a kind of rhythm in the complex disposition of its elements, it is the abstract rhythm of an avant-garde silent film. The writing, as clogged and clotted with qualification, anticipation and flashback as the Jamesian prose about which Sedgwick elsewhere wrote so intently, forms 'a temporality miraculously compressed by the elegancies of language' (*Tendencies*, 184). Not even the discussion of female anal eroticism into which the essay opens out can loosen the stiffly compacted mass of the writing. Perhaps with some residual primness or effort to maintain propriety in the self-disclosure, the thickened white noise of its writing-on-the-spot holds the text back from giving itself over to the rawly masturbatory rhythm to which it elaborately alludes. Perhaps Sedgwick's resistance to giving way, or play, to the actual rhythm which she indirectly evokes in her text is a way of framing the erotic impulse, a framing that simultaneously allows and disallows it – making it possible for the academic to talk dirty, as I am doing now in hands-clean pleasurable proxy, by making the talking the subject of the discourse. Is the device of talking the real English vice, perhaps?

Sedgwick's visualism may in part transmit the influence of 'A Child is Being Beaten', the Freud essay on which 'A Poem is Being Written' is a variation, for Freud's essay also focuses entirely on the complex theatricality of sadomasochism. The girl conducting the fantasy (Freud focuses attention mostly on female fantasies, and indeed the essay is strongly focused around his own daughter Anna, whose sadomasochistic fantasies he analysed, and who subsequently picked up the beat in her own essay on sadomasochism) is in several places, and indeed persons, at once: 'A child is being beaten'; '*My father is beating the child*', '*I am being beaten by my father*'; 'I am probably looking on'.[32]

But though neither Freud (neither of the Freuds) nor Sedgwick pays much attention to the erotic acoustics of sadomasochistic fantasy, they might very well have. The particular play of passivity and dominion that is bound up in the experience of sound is discussed by William Niederland.[33] More specifically, the ritual of counting strokes, or, frequently, forcing the victim to count out the strokes, is a recurrent part of the sadomasochistic repertoire. Counting is part of the ordeal, from German *Urteil*, the base of Old English *adǣlan*, to divide or separate, the doling out of dolour in exact and unalterable amounts. In talionic logic, exactness always participates in the pitiless hardness of exaction. Such a counting rhythm traditionally mechanizes the discourse of the punisher too: '*you-will-nev-er-do-that-a-gain*'. And yet counting, in which the victim is often required to participate, also provides a scansion, which controls, contains and orientates what would otherwise be simple inundation by suffering. Counting can mark an agony of exposure to formless and empty time, which consists blindly and indifferently of one thing after another, each moment a new agony, exactly the same, yet even buildingly more unbearable for that; yet it also provides the capacity for time to be enrhythmed, its indifference given an apprehensible shape and cadence by the strokes, a word that belongs both to beating and timekeeping: 'one, two, three, four, *five*, SIX'.

In all discourse about the relation between regular structure and irregular or unaccountable event, there is an implicit rivalry between the living and the dead. To force someone to count out the numbers of the strokes that are inflicted on them is to force them into the dead condition of a mere number, to force unpredictable life into the painfully regular form of a tattoo, a shape beaten out in air that hardens into visible image incised in the flesh. Such questions of life and death are prominent in the claims made by Simon Jarvis for the cognitive force of poetic music. In his discussion of the melodics of the long poem, Jarvis focuses on the force of the line, which he represents as the primary unit within the design of the long poem:

> The metrical line is the compositional cell of the long poem, before it becomes 'the long poem'; the possibility of recomposition-in-performance, essential to all long poems before they are corralled first into orally standardized and quasi-identically recapitulated, then into written, and finally into printed texts, depends for its possibility upon the formula, a unit which is at once metrical and syntactic and semantic. When all these songs have dried into print, the formula, living repetition as the ever-exploding, ever-generating cell, looks instead like a calque: now sounds, not like the seminal word and tune it is, but like something insufficiently worked over, a dead spot.[34]

A merely formulaic poem is one that has dried into 'rational but helpless quantification'.[35] Jarvis finds in William Collins's 'The Passions: An Ode for Music' an example of a long poem that resists this desiccation into mere number, characterizing the poem as 'a war to the life, in which line must show itself the equal of design, if the whole body is not to become sclerotic'.[36] But the battle between the nervous life of the line and the parched death of the design can only be represented as a war between quantity and quality if one pulls back far enough for the jags and spikes of numeration to be smoothed out into living curves. Turn up the resolution, and the difference between quantity and quality looks like a difference between greater and lesser variation, with the 'compositional cell' as the indispensable unit of quantification:

> the individual line is coloured with the most delicate, the subtlest, instrumentations, not alone with chiasmic

> ornaments of vocalic and consonantal material, but also with interior rhythmic patternings. Now there will be a perfectly antithetical poise between two halves of a line, in which the same tune rings out in both; now the line will bunch all its emphases together in the middle or at the end of the line; now a whole series of lines will run two metres against each other simultaneously, so that a whole passage can be construed either as tens or as sixes; so that the same passage is, as it were, at once epic and lyric.[37]

Try as he might, Jarvis cannot really get number, quantity and measure to stay on the side of death against life. If there is indeed a war to the life against number, it is also fought through and with number.

To bring time under tension – a word that links exertion and music – is to coordinate the ordinal and the cardinal, to fold together counting out and counting up. As one listens to music, one listens out for, or listens in on, the count that one is keeping oneself. Leibniz's unconscious counting involves both recognition of pattern, and conditional projections of those patterns into the open future of what is being listened to. Perhaps all of this is just to characterize the subjunctive mood in which all listening is conducted. The ear is always conducting what it is conducted by, leading what is leading it. This is as true of an utterance as it is of a melody, for every utterance has its distinguishing prosodic profile. Understanding a language or dialect is a matter very largely of tuning into these profiles, learning their landscapes of likelihood, becoming used to keeping time with them.

Just as the human eye looks by default for a face amid a random distribution of visual information, so the ear listens for a voice amid formless noise. What one means by a voice is a particular kind of redundancy, a set of ligatures of the sound that binds it into resonant self-similarity. In this sense, we may say that a voice is simply the personification (*per-sona* = 'through sound') of a rhythm. If a rhythm is the articulation of a flow in recurring patterns, a fracture that is itself refracted back into iteration, then it is number that must register this flexure, since only number allows for this particular kind of segmentation and reordering, this decomposition permitting recomposition. Voice and melody are both probability distributions, precipitates of a calculus.

The world is always between being and number. This is one of the reasons why it is hard to accept formulations like Alain Badiou's regarding the ontology of number as such, let alone his more extravagant claim that number may be Being itself. It does not even seem right to say with Galileo that everything is written in the language of number. We should rather say that nature everywhere tends towards or converges upon number. The Pythagoreans were rattled by the possibility that there could be irrational numbers in nature, but it is in fact whole numbers that are elusive and anomalous. Everywhere there are approximations, distributions, fluctuations around values. Everywhere the real suggests the approach to the rational, but nowhere are the real and the rational absolutely equivalent.

Counting, calculation, and music, the name we give to the pleasure of living out the first two in the listening body, occupy this gap. Michel Serres alludes to Leibniz's remark about the unconscious counting of music in the course of his own evocation of the omnipresence of music in the universe:

> Corporal and formal music, in which, uttering a sort of mute word, the body counts without knowing the numbers. In science, the mind knows that it is counting, it gives the numbers names; music counts by means of unnamed numbers. Inundated by noise, we would be unable without music to enumerate this innumerability.[38]

It is for this reason that music is the intersection for Serres between what he calls the hard – teeming, chaotic, churning materiality – and the soft – form, intelligence, information. Music passes, and itself permits the passage, between the numerous-innumerable and the enumerated.

It is perhaps in this sense that Serres's musical metaphysics construes music as immanent in nature, and lying between form (hard) and number (soft), music as sounded event and music as summable form. But despite appearances, his view is not Pythagorean, for it does not see number as underlying or regulating the universe. Rather nature moves towards number, which arises from it, as a coin being tossed moves towards an absolute 50–50 ratio of heads to tails, without ever settling into absolute invariance. The act of listening to music or, what may come to the same thing, the act of listening musically, occupies a similar space of number in the making, between reality and ratio, arithmetic and rhythm. Music occurs between the *natura naturans* and the *natura naturata*, nature counting out and nature making a reckoning. Music is image and enactment of the oscillating passage between the two.

When I listen to music, what do I hear? Well, I hear 'the music' to be sure, though perhaps I never hear all of the music, and not being able to get my ears round all of it may be part of what that listening involves. Listening is a counting that is not able to take account of everything. But the fact that I must bring myself into a condition of intonation in order to listen means that I listen to something more. Music is a making manifest of listening itself, a listening made musical by lending an ear to itself. Music is the imaginary matter of this listening. What is manifested is what is ordinarily occult in listening, but it is manifested not as a making conscious or as a making explicit, but as a realizing or making actual. Or, it is a making apparent of how much is not apparent to me of how I make myself up as I go along. This kind of listening is not exactly an *exercitium arithmeticae occultum*, an unconscious arithmetical action, or it is not only this. It is an arithmetizing, a making arithmetical, of unconscious action, a realizing of thinking and auditory awareness as a form of quantality. The numerosity of music as a production or 'existing' of consciousness as countable means that listening is not just one mode of consciousness among others, consciousness simply setting itself to the work of listening; it is a way for consciousness to give itself to itself as listening. The Russian philosopher of music A. F. Losev sees music as 'the expression of the life of numbers, a "numeric matter," a meonic–hyletic element that rages inside numeric constructions'.[39] To call this numeric matter 'meonic–hyletic' is to say that it exists between the condition of being nothing, meonic from τό μὴ ὄν, that which is not, and something, hyletic from ὕλη, primal matter (literally, in fact, wood). For Simon Jarvis, music is bound up in the process of binding up the experience of duration, as the actualizing of a sequence of nothings made into a something, into some continuous thing:

> Emphasis cannot but claim that our experience of duration is real. When hours, minutes and seconds drain away in front of us as this sequence of nothings universalized into the measure of life, then outworn iambs, trochees and dactyls carry the promise of a real duration, and, with it, the almost unimaginable promise that our experience might also be for real.[40]

What kind of thing is a listening consciousness? It is consciousness as a mode of self-collecting, in the way, perhaps, in which one is said to 'collect one's thoughts'. Collecting in this manner is founded upon the movement from one to two, as it is described by Fred Kersten:

> The form 'Pair,' or 'the form 'Plurality,' is actualized (or conferred) by virtue of an active collecting (specifically, an active counting or colligating). In the presentation of a pair, we discriminate not only the perceiving, grasping and objectivating 'This' and 'That,' each as self-identical and numerically distinct from one another, but we also can discriminate the active grasping of 'This' and then going on to actively grasp 'That,' still holding 'This' in grip, but still keeping 'This' and 'That' separate. Indeed, the constituting of a pair proves to be the foundation for collecting and counting.[41]

Collecting, like counting, means adding items one by one (they have to be items, or functionally identical units) to a loose, mobile, quasi-totality, without having to hold the whole of the growing sum and all its constituent elements. In counting, letting go is continuous with hanging on, because the number series continues to contain what is at each moment left behind or gone beyond. One need not be or remain conscious of everything one experiences, or experiences of oneself, precisely because one has the relation to oneself of being able to count through. Number, and perhaps only something like number, allows for this kind of coherence-in-dehiscence, this 'numeric matter'. Alluding to Schopenhauer's grandiose rewriting of Leibniz's 'Musica est exercitium arithmeticae occultum nescientis se numerare animi' as 'Musica est exercitium metaphysices occultum nescientis se philosophari animi' – 'Music is an unconscious exercise in metaphysics in which the mind does not know it is philosophizing' – Kersten proposes the further modulation 'Consciousness is a hidden activity which does not know that it is an activity.'[42]

Music is associated with animation, while number is conventionally on the side of death, the mechanical or the inert. But number commutes between the organic and the inorganic, and cannot be conclusively assigned to either. The sheer indifference of number comes from the fact that all numbers are equivalent, or equatable, in that they are made up of units that are exactly the same. Number therefore represents the possibility of a world of absolute indifference. The kind of unconscious counting that is at work in music, that is the work of music, is the effort to capture and neutralize this indifference. But this is in the service of a life that must thereby depend upon and pass through that deathly admixture of indifference that number is.

This may account for some of the pleasure of listening. Listening gives our listening to itself in a way that seems to externalize or automatize it, relieving us of the need to keep hold of ourselves. We do not need to keep the count as long as music is doing the counting, and that counting forms a numeric matter that lets us hear ourselves. It is the pleasure, when it is, of a work that just 'works', a work that does all the work for itself.

9

THE NUMERATE EYE

De Flores, in Middleton and Rowley's *The Changeling*, urges himself on in his lustful pursuit of his lady, Beatrice, with the reflection that she has probably already been unfaithful to her husband:

> for if a woman
> Fly from one point, from him she makes a husband,
> She spreads and mounts then like arithmetic,
> One, ten, a hundred, a thousand, ten thousand,
> Proves in time sutler to an army royal.[1]

Adding is what you do with arithmetic; it is one of the things that arithmetic does. But, in De Flores's imagination, arithmetic itself keeps on adding to itself, since there is ever more arithmetic, which never adds up to anything more precise than 'an army', which is not so much a total as an item that remains to be added up. The addition that arithmetic performs leads to a sum, which is actually a reduction, this the entire mystery of mathematics. The adding of arithmetical operations themselves is an open accumulation, which simply 'spreads and mounts'. There is something lubricious about this very fact of mounting up, that truly lets copulation thrive. De Flores's image prompts us to see two things at once: Beatrice herself, spreading for and mounting her lovers, and the look of the running sum of her past and potential mates as it spills and sprawls across the page. These are the two dimensions of number that have governed this book – the number that has mathematical significance, that performs operations and carries information, and the number that

Arithmetica instructing an abacist at a medieval abacus and an algorist at a counting board, woodcut from Grigor Reisch, *Margarita philosophica* (1535).

exists as pure quantity (and therefore as a kind of quality), number in numbers. There are other appearances suggesting the puzzlement of arithmetic in seventeenth-century drama: Hamlet says of Laertes 'I know to divide him inventorially would dazzle th'arithmetic of memory.'[2] Thersites says of Ajax, stalking up and down, that he 'ruminates like an hostess that hath no arithmetic but her brain to set down her reckoning'.[3] One needs the support of visible space, whether in the form of paper, slate, abacus or screen, to perform arithmetical operations. But if visible space is needed to shape and stabilize reckonings, the very look of numbers and figurings-out can dizzy the eye, as it becomes pure, unreckonable, spreading quantity, accumulating mere numbers of numbers. As Katherine Hunt has observed, the proliferation of handbooks and ready reckoners of all kinds during the sixteenth and seventeenth centuries produced 'numbers in bulk' and 'masses of numbers' that were both unreadable and meant to remain largely unread, since they were designed to furnish answers to particular mathematical problems.[4] Number can be limited by space; but numbers can also haemorrhage through and decompose space. The space of number can provide both orientation and disorientation. This chapter will consider how the visual aspect of numbers operates on both sides of this division.

Numbers are things we say, but they are also, and much more essentially and variously, things we see. There are two ways in which we may be said to see numbers. First of all, we see the symbols of numbers, since numbers are graphical forms, with specific shapes. Numerals are the letters in which mathematics is written. And basic numerical operations require to be performed in visible space. Skills of calculation involve being able to see numerical relationships in a sort of interior space that supplies the space of the piece of paper to which most of us have to have recourse. If children learn to say numbers before they learn to see them, that is probably in large part because their apprehension of number comes through language, and so must pass through the oral into the visual. But numbers seem much more tied to graphical form than words. If there is a language of mathematics, it is to be regarded as primarily a written language.

But there is another sense in which we may be said to see numbers. Most people have to some degree the power of what in 1949 was first called 'subitizing', by which is meant recognizing numbers of items without needing to count them.[5] Typically, most humans are able to do this easily with groups of two or three, and less easily, but still with a fair degree of accuracy, with four, five and six. Thereafter, the process of number recognition may be faster than counting, but not by as much. Though we remain quite good at estimating differences between larger groups of items, this kind of numeration is relative rather than absolute enumeration: we may just have a feeling that one tree has more apples on its branches than another without having much idea of what the total number might be. Unsurprisingly, perhaps, humans share quite a lot of their subitizing capacities with animals.[6]

We may suggest that number and spatial awareness are bound together by more than coincidence. Any kind of manipulation of number, beyond simple operations of merely counting through, seems to require the sense of some space in which the manipulation can take place. One of the huge advantages that came with the arrival in Europe of the zero symbol was that, because it was really just a number-sized chunk of blank space, it allowed calculations to be made directly on as well as in space. The lack of regular consonance between the size of quantities in Roman numerals and the numbers used to represent them made working on and with the numerals directly, as one can work with the beads on an abacus, very difficult. Nevertheless, as Karl Menninger shows, the use of counting boards, reckoning tables and abaci had acquainted human beings with the principle of place-value long before the arrival into Europe of the Indian place-value system in mathematics.[7] As the name of the place-value system announces clearly, it makes value dependent on an item's place in a system. This is the principle that makes difference dependent on a kind of infinite in-difference, since it is impossible to restrict the number of ways in which a number may be defined relative to other numbers – even though this is the only way in which a number can be defined. In articulating the principles of structural linguistics at the beginning of the twentieth century, Ferdinand de Saussure saw that linguistic meaning was dependent on just this kind of place-dependency.

Katherine Hunt and Rebecca Tomlin chart the process by which, during the sixteenth century, 'numbers moved from the counter table to the page, from the material object to the written symbol.'[8] But that symbol never completely dematerialized, for these symbols were still used for doing as well as showing. Double-entry bookkeeping made the book itself into a sort of abacus or calculative apparatus. Rebecca Tomlin has shown how the standing of arithmetic and mercantile accounting was affirmed through the iconography of the books that taught them, like James Peele's *The maner and fourme how to kepe a perfecte reconying* (1553).[9] The arithmetical manuals that began to appear during the sixteenth century, devoted largely to providing instruction in the Hindu–Arab place-value system of calculation, invented a new kind of composite space, that was both an inert framework within which calculations might be made, and a dynamic space, summoned up and scooped out by the act of calculation. In the case of texts like Robert Recorde's *Ground of Artes*, which was written in the form of a dialogue between master and scholar, the page also becomes the mise-en-scène for the arithmetical lesson itself:

> Rather than merely illustrating calculation procedures imagined as taking place elsewhere, such figures come, at different times, to stand in for the fictional space of the staged 'lesson'; for the fictional writing-surface being marked in the course of that lesson; for the thought-space of the Scoler's calculations; for the concrete objects (including sheep and beasts) being quantified; and, in the end, for a hybrid amalgam of all of these that becomes the natural habitat of arithmetical operation itself.[10]

The sense of number requires us to conceive space as divisible into equally sized units. Though we divide time in analogous ways, we in fact always require spatial analogues for these divisions, the time taken for a certain quantity of sand to move through an hourglass, say, or the clock hand to move a certain distance, or the transition between two energy levels of the caesium-133 atom. Number requires the sense of items, of identical chunks of things that may be regarded as numerically identical, or will count as equivalently 'one' of whatever class of thing they are.

The apprehension of space also seems to require some apprehension of relative quantity. Descartes formalized the perception of space with a grid that allowed for any point in visible space to be represented numerically as the triangulation of three coordinates, corresponding to the three dimensions of height, width and depth. This suggested that there is a kind of internalized awareness not only that space must be understood as divisible into quantities, but that it must implicitly be understood as divisible only in a specific number, three, of directions. The very word *dimension* embodies the implicit numbering of space, for it derives from *dimetiri*, to measure out, from *dis-* out, away, asunder, and *metiri*, to measure. Though we may say that we take a measurement in some one dimension, the word embodies some kind of implicit awareness that we are thereby moving away from another dimension, that is, that we are following or creating a bifurcation in space.

These two capacities, one of them applied to numbers of things in visible space, the other to symbols for numbers, and also numbers of symbols, insofar as it may allow us quickly to see that there are three groups of four objects, may often cooperate. Francis Galton investigated the different ways in which individuals visualized numbers and the number line.[11] He reported that about one in thirty males and about one in fifteen females experienced the 'sudden and automatic appearance of a vivid and invariable "Form" in the mental field of view, whenever a numeral is thought of, and in which each numeral has its own definite place'.[12] The Form usually consisted of a spatial schema or array in which numbers were grouped relative to each other, in lines, rows or zigzags. Galton provided illustrations of these Forms along with personal commentaries by their owners. Subsequent experiments have demonstrated other links between numbers and visual space, such as left-right asymmetry; in cultures who write from left to right, larger numbers are associated with the right hand, and smaller numbers with the left.[13] We might understand this as a certain gravitational pull of the spatial in the thinking of number; numbers arise in and from quantities and magnitudes, and are then abstracted from them. The word 'billion' does not take any longer to write or say than the word 'million', and 1,000,000 occupies only slightly more visible space than 1,000, despite signifying a quantity 1,000 times larger. And yet the human propensity to imagine numbers as concrete objects, occupying and distributed through visible space, or indeed themselves forming it, keeps on returning numbers to space.

Indeed, perhaps we may say that, not only are numbers always implicitly spatializable, but visible space is implicitly numerical. Galton was interested not only in the visual dimensions of number, but also in the reverse, the quantification of visual experience. Surprised to find that scientific men of his acquaintance seemed to have very weak powers of mental visualization, he developed a numerical scale to measure the intensity of this faculty and its distribution in society.[14] Galton's two impulses, to number appearance and to visualize number, are combined in the beautiful device he invented to demonstrate how the normal or bell-curve distribution, one of the most familiar and imperious nineteenth-century statistical icons (for once the word is literally correct), comes about in practice. The device, variously known as the quincunx, Galton box or bean-machine, dropped metal beads down a board into which pins had been driven pinball-fashion at regular intervals. The majority of the balls were deflected an equal number of times to left or right, meaning that they filtered through to form a fat hump in the middle of the bottom row; much smaller numbers were deflected consistently to the left or right, meaning that they ended up forming the narrow brim of the bell curve to the extreme right or left.[15]

Our own visual culture is energetically making this latently numerical condition ever more manifest. We may understand the implications of this by considering what it means to think of a sound spatially. This will always involve taking what is compounded (and all sounds are compound, for there can be no sound of anything in itself) and decomposing it in such a way that its elements may be thought of as being set apart or distributed in space as separable elements. Where sound aggregates, vision creates adjacencies. Vision depends always upon a primary act of division, in which individual objects are picked out from their backgrounds. I cannot see the duck and the rabbit simultaneously: I can only oscillate between them. For the same reason, I can never see the whole of a painting, or of any composite object, except by counting it as a kind of one, which is abstracted from its background. One can easily point to similar processes in auditory perception, but insofar as one can, one is pointing to the visual components at work in it. Spatializing or visualizing a sound means turning a chord into something like a score.

It is possible to make a distinction between what may be called mathematical and numerical functions in art. Mathematical art may be characterized as art that is governed by certain mathematical

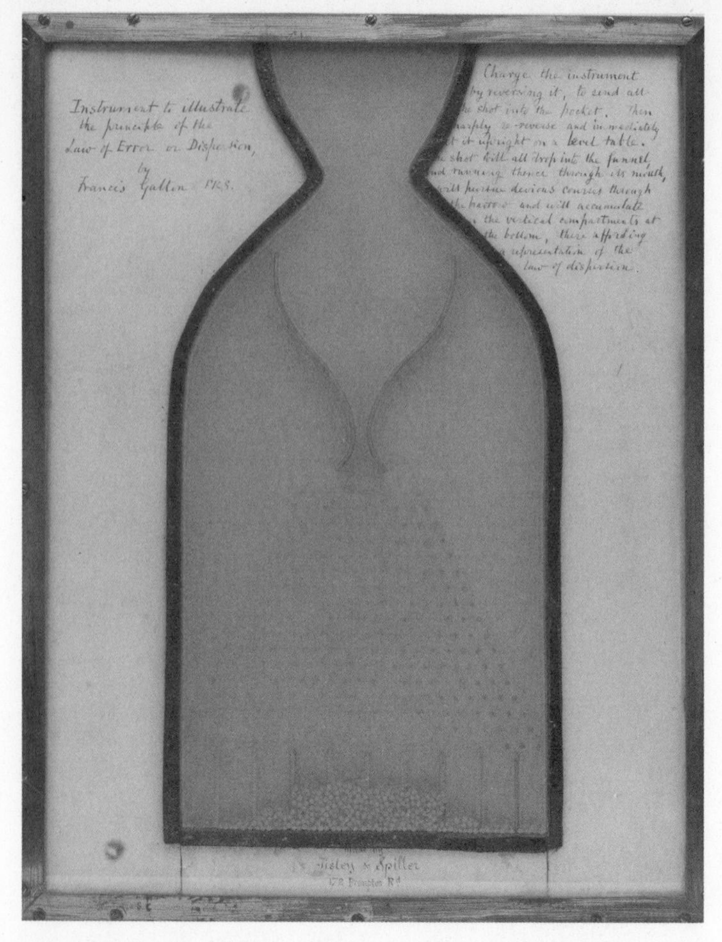

An example of Sir Francis Galton's quincunx, or bean-machine.

principles that may be structural or generative. Art, it is believed, is both governed by and the gateway to certain recurring 'laws' or essential forms, whether these are Platonic solids, the Golden Section or the Fibonacci sequence. Such principles have often provided artists with the occult or cultic authority they crave, after the dwindling of other social or religious guarantees of the worth of art or the standing of the artist. The scorn reserved for 'painting by numbers', the popular practice of the 1950s that has become a phrase that describes mechanically prefabricated kitsch, should not disguise

Jasper Johns, *White Numbers*, 1957.

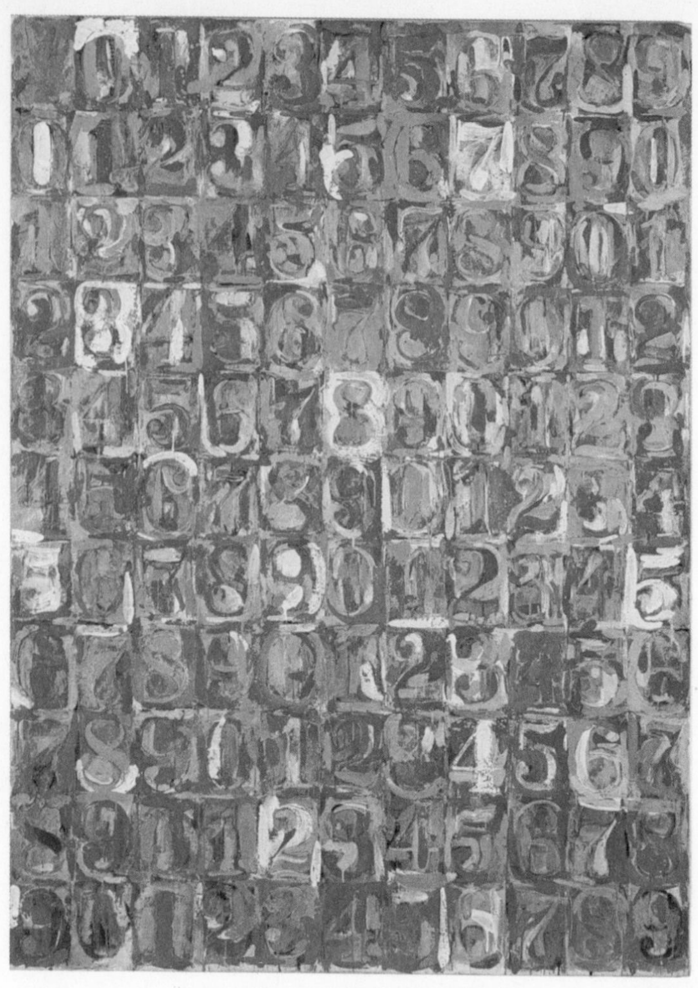

Jasper Johns, *Numbers in Color*, 1958–9.

the fact that it shares a certain structural principle with mathematically construed art and design. Indeed both Leonardo and Michelangelo assigned sections of their paintings to assistants on a numbering principle.[16]

Whereas mathematical art, in the work of artists and writers like Khlebnikov, Duchamp, Malevich, Mondrian, Man Ray and the many others reviewed in Robert Tubbs's *Mathematics in Twentieth-century Literature and Art* (2014), derives its form from mathematical principles, often claiming in the process to provide access to the secrets of natural Form as such, what may be called numerical art operates with numbers as its raw material.[17] Mathematical art tends towards reduction, since it demonstrates the power of a small number of organizing forms or principles; numerical art tends towards expansion, since in it, numbers are plural. In that it tends towards Platonic forms, mathematical art is numinous, rather than numerous. Numerical art is concerned with the visual form of numbers, not their mathematical force (though there may sometimes, and perhaps usually, be a certain play between the two kinds of idea). What matters here is not primarily the signification of the numbers, except insofar as they may signify the elementary forms of numericity, the simple fact of being numbers.

The visual forms of numbers have been employed as ways of importing immediate and unfalsified bits of reality into otherwise abstract painting, for example in the collages of Braque and Schwitters. Probably, though, no painter has made such frequent and systematic use of number-forms as Jasper Johns. Johns first began painting numbers in the late 1950s, during a period in which he had been painting a series of ordinary objects, such as flags, maps, targets and letters. The choice of such objects, along with Johns's habit of painting them repeatedly, has been understood as motivated by a desire to 'drain subjectivity from subject matter', muting the expressive drama that could attach to abstract painting. In fact, of all the commonplace visual forms that feature in Johns's work, numbers are by some way the most numerous.[18] He has returned to them throughout his career, producing around 180 works in different media from 1955 onwards, including, in recent years, sculptures cast in bronze and silver.

Johns's painting of, as opposed to painting by, numbers creates a striking antinomy. On the one hand the familiarity of numbers, combined with the fact that Johns gives them easily recognizable generic forms, often using stencils, rather than elaborating them like historiated initials in illuminated manuscripts, seems to rule out aesthetic responses, making the viewer wonder embarrassedly or irritatedly what kind of contemplative opportunity is being proposed for them. In his first sequence of number-paintings, representations of single numbers he called 'Figures', Johns even took steps to counter any suggestion of a systematic project, explaining that 'I didn't work on them in any order and I deliberately didn't do them all, so that there wouldn't be implied that relationship of moving through things.'[19]

And yet the very ordinariness of these number-forms also serves to highlight by contrast the quality that art historians like to call 'painterliness'. Roberta Bernstein asserts that 'Johns rewards the eye by the sensuous handling of palpable textures of his surfaces' and Carolyn Lanchner writes with uncurtailed relish of *White Numbers* that the 'messily meticulous surface alternately seems as delicious as cookie dough and as fragile as a veil drawn over a dream – a curious fate for the numerals of decimal narration'.[20] The very dullness and affectless aridity of number is precisely what gives appetite and excitement to the act of painting them, an act that ends up intensifying rather than depleting the sense of visual significance. As Charlotte Buel Johnson says of Johns's *Numbers in Color*, 'the numbers have nothing to do with telephoning. They have nothing to do with banking or arithmetic. The numbers in the picture are part of a design. They have become something new. They are just shapes and colors repeated and repeated.'[21] It is the very unartistic banality of numbers that is supposed to foreground the strange, even in its way heroic, self-jeopardizing of the work of art that takes them as subject. This can lead to the conclusion that the aim and outcome of these paintings is an interrogation of the conditions of perception, as in Roberta Bernstein's judgement:

> Johns's Numbers are unique in their singular focus on the number itself as a formal and conceptual entity. The artist's

> strategy is to neither [sic] avoid meaning nor hide it. Instead his Numbers, like his other commonplace objects and signs, become vehicles for examining the process of perception and the fluidity of meaning. Johns's presentation of numbers as uncertain signs is first and foremost aimed at stimulating the viewer to awaken the eye and mind to experience what is most familiar in a new way.[22]

This is a judgement – a wearily conventional one – about what art does, not about what the visual force of number-forms might be. But, armed with the sovereign principle that the most interesting things about art rarely in fact have anything to do with art, we might see what there is to say, to say what there is to see, about that force.

Michael Crichton observes the oddity in Johns's number-paintings of 'the idea of representing what is already an abstraction. A flag is an abstraction, though most people think of it as a piece of multi-colored cloth. But numbers exist only as intellectual constructs, and to give them form in a painting is to challenge immediately our ideas of representation and abstraction.'[23] But numbers are abstract, not in the sense that they can never have any reference to the world, but in the sense that their reference is not limited to any specific objects in the world. There is no number that could not in principle designate a certain number of some thing or other, even if the number of things that it could so designate is in fact numberless. Not only this, numbers are themselves countable things, as emphasized in Johns's paintings, which suggest that their forms have been physically stamped into the material of the painting, like a seal or a brand. In late works like *Numbers* cast in bronze of 2005, or the 0–9 cast in silver of 2008, this blocky, monumental materiality is made conspicuous. Numbers belong, I have just been saying, to the order of adjacency, to a space the principal quality of which is that things lie next to each other. Johns's numbers, or more exactly his Numbers, the sequence of repeating grid-arrangements of the numerals from 0 to 9, are packed tightly together. Rather than floating in space, they form and entirely fill the space they occupy. Their forms may remind us of children's building blocks, or of the stencilled numbers on packing crates; in either case, there is the suggestion of the block of

Jasper Johns, *Gray Alphabets*, 1956.

Jasper Johns, *Map*, 1961.

type the purpose of which is to print off signs. It is not surprising that Johns should have arrived at his number grid through painting of other gridded forms, notably the alphabetic letters of *Gray Alphabets* (1956). But his painting of cartographic forms, as in the United States map rendered in *Map* (1961), indicates that it is not the grid arrangement that is important so much as the idea of a packed space, of abutting and adjacency leaving no more gaps than a wall of bricks. Johns shows us a space that is not abstractly and arbitrarily overlaid with the abstraction of number but rather filled with it.

And it is this principle of saturation that governs the series of paintings known as *0 through 9*, as opposed to the almost identically named 0–9 sequence. In these paintings and drawings, the principle of adjacency required and sustained by the grid is apparently violated, since the numbers are here superimposed rather than jammed alongside each other, creating a deep, dense, entangled, scarcely legible space, as though Johns were rendering simultaneously the sequence of numbers that used to flash into view at the beginning of a reel of film. The clue seems to be given by the word 'through' in the title of this series. Each of the numbers is seen through a lattice formed by the other numbers, creating a pattern of visual interference which makes it hard to make out any particular number in its entirety. But this is still a packed space, a space that is full of

Jasper Johns, *0 through 9*, 1961.

numbers, in which even the interstices between the numbers are part of the space enclosed by some other number and so will allow room for nothing but number.

Johns here gives us images of number as the very plenitude of space, number as the principle that opens and occupies every void. At times, it might appear as though his aim is in fact to obliterate all visible space – obnumerate it, we might almost say – by swamping it with number. Yet it is the visible forms of numbers, still just discernible and therefore holding the complete indifferentiation of whiteout or blackout at bay, that make space apprehensible as such. This is made manifest in the painting *Thermometer* (1960), in which a single white line runs down the middle of the painting, reminiscent of a column of mercury, with two black dots perhaps marking the positions of freezing point and boiling point. The rest of the picture is almost an unrelieved black, apart from the curves of what appear to be numerical calibrations emerging from the murk. To be able to distinguish is to be able to number, and to be able to number is what gives space its form.

In 1965 the French–Polish painter Roman Opałka made a start, with a stroke of white paint signifying the number 1, on what would be the single work that would occupy him for the rest of his life. He started at the top left-hand corner and continued painting the sequence of numbers, 2, 3, 4 and so on, from left to right in closely compressed lines. Each time he completed a canvas, he began the next one with the number immediately following the last number painted on the previous canvas. So, we may say with equal justice either that Opałka painted a large number of number-paintings, or that he only painted one, in a number of instalments. His first canvases used white paint on a black background, but, during the 1970s, Opałka took the decision to begin lightening the background, adding 1 per cent more white to the pigment with each successive canvas. He calculated that by the time he reached the number 7777777 (Opałka never employed commas) the numbers would cease to be visible against their background. Later, Opałka began to record himself enunciating the numbers as he painted them, and to take photographs of himself at the end of each day's work, thus, it seems, creating an absolute continuity between his life's work and

Roman Opałka, detail from 1965/1–∞, *Detail 1-35327*, 1965.

his actual life. Opałka's days were, more literally than any other's, numbered. When he died in 2011 at the age of 79, he, or the count, had reached 5607249.

Karlyn de Jongh writes confidently that Opałka's practice is 'a radical program through which he seeks to portray the passage of time'.[24] Opałka himself has called his work 'eine Progression, die die Zeit sowohl dokumentiert als auch definiert' – 'a progression which documents as well as defines time' – and a 'Visualisierung der Zeit' – 'visualization of time'.[25] But, although the unbroken series of numbers might give the impression of representing pure duration, it can do so only in a very abstract and approximate sense. For the steady sequence of the numbers is an idealization of what in actual fact was a necessarily discontinuous and episodic enterprise, even if Opałka did succeed in sustaining it until the end of his life. The clock of Opałka's painting practice ticks spasmodically and irregularly. Whatever the interest of what Opałka is doing, it is nothing as banal or showy as this kind of temporal allegory. In fact, if Opałka has any design on time, it is surely rather to show its unshowability, its unthematizability.

The idea of painting something as theatrical or thematic as the 'passage of time' is in fact a derivative, perhaps even a deflection from the primary activity of which these paintings is composed, of painting numbers in sequence. Every effort to round this activity up into some larger (which is also to say smaller) purpose wilfully evades and effaces its actuality, which lies in its failure to add up to anything. This is even true of the solemn purposes to which Opałka has allowed his work to be dedicated, as when he built an installation from the sounds of his counting in the Delme synagogue in Lorraine in 1997:

> As with every synagogue, the Delme synagogue shouts out the memory of the victims of the Shoa; the victims whose names are crossed out and 'who were no more than numbers' . . . Every presentation which is more than just the memory of the number – including all the horror of a past still young and for ever unforgettable – has seemed to me out of place. As for my part, I wanted to give another sense

> to the numbers, especially in this place. I have been painting them and enunciating them for so many years, one after the other, and thus I have been creating a dynamic uniqueness of every being – similar to a prayer just told once like a life only lived once. So my work turns into a testimony, into a path, into some patient determination to follow them towards the light.[26]

Whatever Opałka's apparent and even declared purposes, what he in fact devoted his life to was the painting of numbers in sequence, from one onwards. One might call this an act of counting, except that nothing is here being counted but numbers themselves. This is the process of what it is, not any kind of determination to be 'in the moment' or anywhere else; it is in fact a determination to avoid having been anything at all or in particular. Opałka said in 2009 'How can you understand a thing as stupid as our existence? Maybe that sounds too brutal, but this existence makes no sense; it is nonsense. And this nonsense is my work.'[27] And the work's way of avoiding or denying ontology is through pure number, or, rather, what might seem to stand on the other side from pure number, the unrelieved act of numbering.

There are many epiphenomena attending this enterprise. One is that the white paint used to inscribe the numbers fades as the line proceeds, so that the brush must be loaded with more paint, usually after no more than ten numbers, and sometimes after as few as three. From any kind of distance, the microrhythms formed by these fadings and replenishments give a visible texture to the surface of the painting, like the ruffling of a surface of water, or the layerings of slate or tree bark, or the textures of squamous skin.

Another syncopation is provided by the fact that there are frequent mistakes. Indeed, there are miscounts in the very first few lines of the first painting in the series: there is no 416 between 415 and 417, and the sequence jumps from 425 to 456 missing all the numbers in between. Later on there is no 3811 or 4004, but two 3999s, two 4006s and two 4552s. Opałka seems not to have sought to correct these mistakes, though the practice of saying the numbers out loud he later developed may have been a way of guarding against them. Among other things, this makes it clear that Opałka's paintings are not, as one might otherwise assume, self-enumerating. The only way to establish whether, by the time he painted the number 5607249, Opałka had in fact painted 5607249 numbers would be – amazingly, appallingly – to count them.

Far from portraying the pure passage of time, Opałka's paintings show just what they set out to show, namely the visibility, or, more precisely, the making visible, of number. In this, the effect if not also the purpose of the sequence is not to gratify or congratulate the eye, by giving it an image of some idea, but to baffle and to nauseate it, allowing no vantage point or point of rest. Almost all that can be made out in, or of, Opałka's work are huge, though not innumerable, numbers of numbers. There is no way to get any of the pictures in focus. Like numbers themselves, they are all different, yet different in precisely the same way. The effect of twinning the images of numbers with the self-portraits of Opałka's steadily ageing face is to suggest the merger of face and figure (the latter in fact derives from Latin *figura*, a face), as though the face itself were to be offered as a sort of graph of the approach to decay and extinction, at once the contour of an impassioned life, full of life and joy, and the indifference of the process by which it is composed and decomposed through time, as time. Opałka's face comes to have the look of number, as his life had number's shape and tempo.

Opałka's practice points up the sentimentalism of the practice of 'numberism' of an artist like Sienna Morris. Though she uses numbers as constituent marks for the figurative forms she designs, the numbers themselves are given an emblematic appropriateness to the artistic subject – so, for example, her drawing of the human heart is done entirely with equations used by cardiologists to determine the health of a heart, and, in her portrait of Einstein, the equations for relativity can be seen curling in the smoke from his pipe. But this is a purely arbitrary association: insofar as they are not immediately visible, the numbers into which the visual forms may be decomposed need not be cardiac or Einsteinian equations, or indeed equations at all. Opałka's work, by contrast, offers us no defence or retreat into anecdotal or philosophical significance from the exposure to pure, indecipherable ciphering.

No doubt the popularity of works which appear to be made of numbers in itself constitutes a sort of general allegory, since we seem to live in a world that seems more and more to be made of numbers, or equivalently numerical coded forms, rather than of pure matter. Such pictures may serve to image the sense that we can never now go down far enough in nature to arrive at the condition of pure matter, beneath the condition of number.

The difference between these two modes of visualization is paralleled by the two near-contradictory ways in which the word *graph* is used in mathematics. First of all, and most familiarly, graphs are visualizations of numbers, typically of functions that produce variable results over time. But graphs also refer to mathematical problems that are visual in their nature, that is to say, largely problems of topology, such as Euler's bridge problem or the four-colour problem. The first kind of graph moves from number to image: the second moves from image to number. In the first kind of graph, N>I, since the image represents a reduction of number to image, in the second, I>N, since in it visual form is reduced to number. The two kinds of graph encode what have often been thought to be the two essential forms of mathematics: geometry, which may be practised without recourse to measurement, and algebra, which renders spatial relations in numerical form.

NUMBERS IN PUBLIC

Roman numerals belong to public time, the time of proclamations and memorials. Philip Larkin's evocation in his poem 'MCMXIV' of the world of 'innocence' that was swept away by the First World War depends upon the use of Roman numerals in its title. The numerals impart to the poem itself a kind of 'look', which is internally reduplicated within it by the evocation of various kinds of inscription, words that are not so much read and instantaneously seen, as in a photographic exposure: 'the bleached/ Established names on the sunblinds'; 'The tin advertisements/ For cocoa and twist'; 'The place-names all hazed over/ With flowering grasses'.[28] In a certain sense, this direct visibility is part of the innocence of meaning that the poem names. There is an implied interpenetration of natural form and number in the evolution of the act of political and economic accountancy that gridded the country after the Norman Conquest, in the 'fields/ Shadowing Domesday lines/ Under wheat's restless silence'.[29] Domesday is the name for a certain kind of stopping of the clock for the purposes of reckoning and the mapping of quantities. A world of inscriptions of this kind is a world of naturalized number. It is part of the way in which things last, but is also exposed to what is about to happen, hinted at in the way in which the word 'bleached' seems instantly, and paradoxically, to decay into 'Established'. The reckoning that is in store will be made with different kinds of numbers, in the thousands and millions of dead and injured of the First World War.

But the world of 1914 was a world in which the compoundings of image and number had already hugely diversified, for it was a world in which the graphical representation of variable quantities had already become commonplace. Just glancing at mathematical books over the course of 500 years in Europe shows three phases. In the 1500s, mathematical books were presented as unforgiving blocks of text, broken up only occasionally by diagrams. By the eighteenth century, numbers and mathematical symbols occupied much more of the page, steadily displacing the text. The image of the mathematician's blackboard completely covered with equations expresses this phase of mathematical work. But in the third phase, abstract symbols have given way to concrete visualizations, which are at once symbolic and iconic, at once an abstract formalization through numbers, and a concrete visualization of those numbers in terms of quantitative magnitudes and relations.

There is one spatial form that is more expressive of numeric space than any other. The grid, formed of parallel vertical and horizontal lines intersecting at right angles, has for centuries been the most effective way not only to picture space but to govern it through numbering operations. In *Through the Looking-Glass*, Lewis Carroll invites us to imagine a rationalized landscape as laid out like a chessboard, but this is what all maps and charts had done since Descartes devised his coordinate system in 1637. By the later nineteenth century the Cartesian grid had become visible in physical space, as cities began to be built on this plan. Although the grid rapidly came to be thought of as characteristic of modern cities, it had in fact been employed in many ancient cities. Hippodamus of Miletus designed many cities in Greece on the grid plan and the design was carried far to the East by Alexander the Great. That the grid was intimately tied to number, and the control through abstraction that allowed, is indicated clearly by the name *centuriation*, which was given to the Roman grid system. As Hannah B. Higgins has shown, the grid form of urban design conjoins the organizing function of two forms – the brick and the cuneiform tablet – drawing form into information.[30] The grid borrows from and displays the two essential principles of the number system: firstly, the identity of units, each cell of the grid being regarded as an equivalent numerical element, and, secondly, ordinality, the ordering, or orderability, of these equivalent elements in a fixed series that allows for location through counting.

This means that the grid is a uniform space. Any portion of the grid is completely interchangeable with any other portion, even though numerical specification also allows for the unique identification of individual locations within it. In fact, the two principles are codependent; there can only be unique identifiers within a system in which every unit is in principle entirely exchangeable with any other. Such uniformity implies a certain kind of executive power that is capable of maintaining singleness of intention. This is why the grid form of town planning is associated with dictatorship, absolute power or great wealth, since the impediments to be overcome in reducing an historically irregular space to a spatially homogeneous one, laid out to view as though for a single shared eye, are considerable.

The space of the city can also recall other kinds of network, which can occasionally come into a sort of visibility, as in the description of 'San Narciso' in Thomas Pynchon's *The Crying of Lot 49*:

> She looked down a slope, needing to squint for the sunlight, onto a vast sprawl of houses which had grown up all together, like a well-tended crop, from the dull brown earth; and she thought of the time she'd opened a transistor radio to replace a battery and seen her first printed circuit. The ordered swirl of houses and streets, from this high angle,

> sprang at her now with the same unexpected, astonishing clarity as the circuit card had.
>
> Though she knew even less about radios than about Southern Californians, there were to both outward patterns a hieroglyphic sense of concealed meaning, of an intent to communicate. There'd seemed no limit to what the printed circuit could have told her (if she had tried to find out); so in her first minute of San Narciso, a revelation also trembled just past the threshold of her understanding.[31]

Grids allow for space to be numerized because grids are also the characteristic forms employed for mechanisms of calculation – like the abacus, the loom (which, via the punched-card system invented to govern Jacquard's loom, gave rise to the computer program but can be seen as a kind of computational device in itself) and the cellular arrangements employed for paper calculations.[32] The grid shares many features with the table, and forms which borrow its principles, like musical scores and entablatures. Michel Foucault has observed the ways in which the flat surface of the table allowed for a structure of knowledge that depended on the principle of ordering things into classes. The table is always a version of the *tabula* 'that enables thought to operate upon the entities of our world, to put them in order, to divide them into classes, to group them according to names that designate their similarities and their differences – the table upon which, since the beginning of time, language has intersected space'.[33] This intersection of space, or space of intersection, is also where language is intersected by number.

For this reason, the table always exists between the actual and the abstract. The table allows for different items to be laid out and held open to view alongside each other. A physical table is always the synecdoche or place-holder for the more abstract space of relation as such, which is to say, precisely, a space in which places may be occupied, taken and changed. It is for this reason that a table must always itself be stable, in order that it can itself encompass variations. The table comes into being whenever the ground is doubled, the ground lifted up for the convenience (literally the 'going-together') of eye and hand. The table on which things are arrayed is always the

Tavola 33. הדה

G	1	18	8	7	1	[illegible]
F	2	1	3	47	2	[illegible]
M	5	4	1	2	3	[illegible]
A	5	1	8	1	4	[illegible]
M	4	81	1	0	2	[illegible]
G	2	17	18	18	0	[illegible]
L	3	2	4	0	2	[illegible]
A	6	3	1	0	17	[illegible]
S	1	13	40	17	1	[illegible]
O	4	40	17	2	0	[illegible]
N	8	7	2	13	43	[illegible]
D	8	2	4	22	0	[illegible]

Tavola 35. ההל

G	5	45	0	2	8	ה
F	4	1	15	0	2	ד
M	1	88	3	0	7	ח
A	9	70	10	1	2	ב
M	8	17	8	12	8	ה
G	6	2	22	1	1	ד
L	3	5	14	2	0	ל
A	3	3	4	74	1	ה
S	4	2	4	15	5	ב
O	2	17	70	1	0	ב
N	1	4	4	8	0	ה
D	8	18	8	1	0	ע

Tavola 34. חדה

G	9	13	4	40	0	[illegible]
F	9	90	0	1	2	[illegible]
M	8	1	2	3	49	[illegible]
A	5	14	47	12	1	[illegible]
M	6	7	2	14	61	[illegible]
G	2	1	18	8	7	[illegible]
L	8	2	41	4	0	[illegible]
A	8	4	1	2	17	[illegible]
S	8	10	13	37	0	[illegible]
O	4	1	10	10	1	[illegible]
N	4	2	22	0	4	[illegible]
D	9	32	4	1	5	[illegible]

Tavola 36 הדהלה

G	1	8	8	14	6	ה
F	1	2	4	18	7	ה
M	9	41	5	6	5	ב
A	8	17	1	2	10	ה
M	4	13	0	1	14	ד
G	1	12	47	1	4	ד
L	1	13	11	1	14	ה
A	7	12	4	2	13	ד
S	8	18	80	1	0	ב
O	7	1	18	8	4	ס
N	4	4	7	1	1	ה
D	2	9	0	8	0	ד

Four numerical calculation tables based upon the work of Johannes Reuchlin, using Hebrew letters, right-hand column, and numbers, c. 1838, pen and ink.

James Murray's filing system for the OED.

space between those things, the *super* always an *inter*. This remains the case even with the infinite desktop of the computer, which still requires the material support of the screen. As the meeting place of the corporeal and the cerebral, tables are used for eating as well as for writing and for playing – etymologically, a tavern is a place of tables. Lautréamont famously evoked the beauty of the meeting of an umbrella and a sewing machine on an operating table, but all tables are operating tables really, in that they all allow operations, of work and play.[34] To eat at a table is to mediate the raw and the cooked, to make the consumption of food into a social action, which takes place in a visible space in which I must not only know, and see, my place, but also recognize that I may change places with anybody else, whether in secular company or religious communion. The dissection table or operating table, always, as in Hogarth's *Reward of Cruelty*, depicting the public dissection of a hanged criminal, shows us a kind of grisly cognitive feast as it displays for inspection the viscera of the human body laid out upon it, for visual consumption by the other bodies gathered round. The protocols of dissection and surgery remind us of the careful ordering of elements in visible space required by Leviticus in the sacrifice made at another kind of table, the tabernacle, or altar, so named because it lifts up what is laid on it. In the sacred diagram of the sacrifice, as Mary Douglas has shown, 'position is everything'.[35] The table is a place of play and performance, a place in which place values can be put in play. This makes the board and 'the boards' of the theatre equivalent. To 'turn the tables' on someone literally means to change one player's position for another, 'tables' being the name for backgammon, because it is played with hinged boards. The chequerboard pattern of the chessboard and the tablecloth images this reversibility, which has been, as it were, folded into the space of the table itself to become part of its texture. In all of this, the equivalence of spaces and the elements distributed between them makes of the table a countable space, a space open to number. Commensality is commensurality.

The grid also has a central significance in modern art. Grid arrangements seem ubiquitous and tenaciously long-lived, featuring in the work of Mondrian, Malevich, Klee, Reinhardt, Martin, Warhol, Andre and, as we have seen, Jasper Johns. The point of the grid, asserts Rosalind Krauss, is to proclaim the modernist autonomy of art. The grids employed to determine perspective in the work of Leonardo and Dürer were not really grids in the modern sense, because they were devised to enable a mapping of the real onto the two-dimensional surface of the artwork. The modernist grid, by contrast, displays the fact of displaying nothing but its own ordering schema: 'In the spatial sense, the grid states the autonomy of the realm of art. Flattened, geometricized, ordered, it is anti-natural, anti-mimetic, anti-real. It is what art looks like when it turns its back on nature.'[36] The painting organized like a grid is not imitating the world, but indicating its own powers of visual organization.

This is, however, a surpassingly strange claim either for art, or for art criticism on its behalf, to make. For the very thing that makes the grid count as a regulatory space is the fact that it recalls all of the many kinds of spatial arrangement of this kind that have become familiar, in maps, plans, diagrams, games, puzzles and designs of every kind. It may be that the modernist grid subjects the real to the 'overall regularity of its organization', but it can scarcely be that 'this is the result not of imitation, but of aesthetic decree'.[37] The art of the grid asserts its absolute autonomy by mimicking something else – geometrically regularized space. That is, art asserts its autonomy through heteronomy, demonstrating its independence through its dependence on something else (given the ubiquity of grid arrangements in modern life, one might say its dependence on almost everything else). Grids need not themselves be numbered, but they belong to the large class of objects that now appear available for numerical operations. Krauss acknowledges this duality in her observations on the two kinds of grid that feature in modernist art, in one of which the grid extends, like the number line, infinitely outwards from the frame of the painting, the other of which 'is an introjection of the boundaries of the world into the interior of the image: it is a mapping of the space inside the frame onto itself'.[38] But the two operations are really equivalent: the modernist grid can only turn in on itself by implicating itself in a general economy of gridded spaces. The look of a modernist grid is in fact the dominant form taken by the look of number.

The coordinate space of the grid is typically without depth (though this in fact depends on the number of dimensions that the coordinate space includes). For this reason, the Cartesian grid is a route to the mathematical projection of *n*-dimensional spaces, through the simple addition of axes on which to add coordinates. So one might imagine, beyond the horizontal axis *x*, the vertical axis *y* and the *z* axis of depth at right angles to both, limitless other axes, at different angles of incidence. Once again, we may see – or almost seem to see – the innumerable arising from the enumerable.

Peter Greenaway has employed the flat, grid-like spatiality of the picture plane to complicate the construction of and response to the moving image. Greenaway has spoken of his attraction to the simple play of horizontals and verticals, remarking, in an interview with John Petrakis in 1997:

> I'm looking all the time for alternatives to storytelling. My films are very much based on horizontals and verticals. It's

> a grid situation. Also lists, number counts and alphabetical counts. Not that I believe intrinsically in any particular magic in these systems, but they are well-defined, well-wrought systems of organization.[39]

Visual arrangements are put into counterpoint with narrative, which Greenaway regards with a certain tolerant suspicion, being unwilling to abandon it entirely, but remaining wary of its factitious satisfactions. *Drowning by Numbers* is the Greenaway film in which the conflict between the ongoing time of narrative and the screen as a space of merely numerical unfolding is most clearly marked. Speaking with Hartmut Buchholz and Uwe Kuenzel in 1988, Greenaway evoked the interplay between the inner core of narrative and the outer armature of number:

> I use two skeletons – one builds the core and the other creates the outer form. If from one point of view all this sounds trivial – well, it most certainly is. But naturally I incorporate another idea: I believe that we have very narrow margins to express our 'free will.' Ostensibly, we are capable of making decisions, but these decisions are really very limited. The film should exhibit this by having its story as inner skeleton embed itself in an outer one which reveals this limitation – something like fate, though I don't like the word. Our lives, after all, are circumscribed by conditions over which we have no control – our surroundings, the climate, our personal contingent relationships. For me, the mathematical structures signify those boundaries that constrain us.[40]

Drowning by Numbers is an intensely if also playfully English film, with its references to game playing and rule giving, its hints of Lewis Carroll, Agatha Christie and detective story. But, like many other Greenaway films, it seems to inhabit the intersection between the corporeal and the abstract – sexuality and pattern, decomposition and composition (hence the importance of insects, rationally segmented creatures who are associated with decay), the desire for death and the death of desire. The two coordinate systems that seem to be placed perpendicular to each other are sex and number, approximating to the corporeal and the cerebral. The film's narrative is concerned exclusively with sex, as the most inclusive form of bodily desire. A sequence of three men are drowned for reasons of sexual inadequacy, an impotence that seems to be associated with their obsession with pattern, number and game. The three murderers, all named Cissie Colpitts, ensure that they escape retribution by promising sexual favours to the coroner Madgett, and seem to get away with it. At the end of the film, Madgett sits, naked, in a scuttled rowboat, possibly about to become the fourth victim of drowning. Meanwhile, his son Smut has hanged himself, following his own self-circumcision. If this act seems like an irruption into the ludic dream of the film of something like the violence of the real, even this reality is conveyed by and contained with a game structure, as Smut himself narrates his own suicide: 'The object of this game is to dare to fall with a noose around your neck from a place sufficiently off the ground such that a fall will hang you. The object of the game is to punish those who have caused great unhappiness by their selfish actions. This is the best game of all because the winner is also the loser and the judge's decision is always final.'[41] Games hold violence at bay, but they also, as we are reminded throughout the film, lead to and are led by violence; Madgett's obsession is with a history of injuries incurred in sport, including this episode: '1931? Chapman Ridger? Australia? A blow on the chest (*He thumps his chest*) . . . Hits 51 runs . . . then has heart palpitations for twelve hours – a cracked rib enters the lungs . . . coughing blood, dies the next day in bed with a blonde surfer called Adelaine.'[42]

Finality is promised by number, even as numbers, which go on for ever, also defy finality, making for variability, repetition and reversal. In the end, it is hard to know where the film stands in relation to the work of number – or even what we might mean by 'the film'. Is it the narrative that is conveyed by the film, a narrative that has everything to do with counting, or is it the mere succession of its numbered frames, which in this case have actually been numbered for us, in the various signposts that have been strewn through the film, from the number 1 painted on a tree trunk in the opening moments of the film, through numbers made more or less conspicuous all the way through to the 100 painted on the sinking rowboat in which Madgett sits? Greenaway notes the evasive, anaesthetic powers of counting, which enables us to evade death, even as that evasion is itself deathly:

> Counting is like taking aspirin – it numbs the senses and protects the counter from reality. Counting makes even hideous events bearable as simply more of the same – the counting of wedding rings, spectacles teeth and bodies dissociates them from their context, to make the ultimate obscene blasphemy of bureaucratic insensitivity. Engage the mind with numbing recitation to make it empty of reaction.[43]

But it is not as easy as might be thought to thematize number, or give it its secure place in relation to what we see played out in the narrative or theme of the film. The men count, obsessively, as part of their evasive or suspensive game playing; but so do the women, as part of their serious work of murder. The skipping girl who opens the film counts the uncountable through proxy (her chant names a hundred stars and stops because 'a hundred is enough. Once you have counted one hundred, all other hundreds are the same'), while Smut counts the accidents of violent deaths.[44] In the end, the apparent significance of the struggle between male and female, and the corporeal and the abstract that it approximates, is itself dissolved in the universal solution that is simple numbering, which both runs through all the action and is orthogonal to it. It is as though the film were like Carroll's Red King, muttering to itself, and to us, 'Important – unimportant.'[45] It is a film that is as delicious to think and write about watching as it is almost intolerable to watch, in its inhuman ticking off of the seconds and minutes, and the remorseless and indifferent equivalence of beauty and violence. Seeing number and seeing numerically, which can so often be a form of visual triumph over the unassimilable phenomenality of appearance, is here shown or seen as a kind of pathology, a pathology that seems to go with the grain of vision itself. The only remission from the order of number seems to be in the rhapsodic registering through the film of the unpredictable and incalculable movements of wind, water and flame – the number 1 was painted on a tree that had actually been toppled by the freak hurricane that blew across southern England in October 1987, and was artificially set upright for the film.[46] As he edited the film, Greenaway began to decompose it into ever smaller constituents, opening up a Borgesian infinity of numerable and permutable elements:

> Nearly every day for three weeks, I watched the film through the many print-stages – each one moving closer to a fully graded print that would satisfy. I enjoyed watching every one and – sometimes admittedly compensating for its familiarity – I began to see the film in different ways. One of those ways was by observing the small gestures. Sometimes I watched the film entirely through its small gestures.
>
> The green eye of the corpse that swings on the post in the light of the lighthouse . . . the plop of the invisible frog that jumps in the pond in the night-garden behind the titles . . . the light on Nancy's breasts in the brown and orange atmosphere of the apple-garden . . . the rumble of the apples as they are tipped from the tin-tub under the apple trees.[47]

Though Greenaway has been at pains to emphasize the ways in which numerical frameworks constrain, number also seems to lead to proliferation. As so often, numeration seems to encourage ever greater numbers of numbers. As well as giving visual signals of the count from one to a hundred through the film, Greenaway buried significant numbers in it. The bedroom inhabited by coroner Madgett contains a hundred objects starting with the letter M, while his son Smut has a hundred objects beginning with the letter S.[48] In one scene, Bruegel's painting *Children's Games* is shown resting on an easel by Madgett's bed. Greenaway observes of the painting that 'every game – 84 of them – can be identified and most of them are still played in some variation today.'[49]

Drowning by Numbers seems to have been the stimulus for a series of other Greenaway projects that would continue the work of counting, onwards and outwards from the closed 1–100 series that counts the film off and out. The film spawned a plan for an eight-hour TV serial to be called *Fear of Drowning*, which Greenaway described in an interview with Stuart Morgan in 1983. The film was to trace the life of Cissie Colpitts up to the age of eighteen, on the date that the Lumière brothers patented the first cine camera, making her life parallel the history of the cinema:

> It shows that she inherited both her gameplaying and her terror of drowning from her father, a man called Cribb. Every episode will contain a different game. The first, learned from a shipwrecked Italian sailor, Cribb plays on a beach to determine his daughter's future. It involves drawing squares, each of which is your destiny. You play hopscotch and throw rocks. Where they land indicates patterns of behaviour. Another, the Lobster Quadrille, is an obstacle race relating to all the fears sailors have of the sea: deep chasms in the China Sea, the aurora borealis, the Sargasso Sea, the Strait of Magellan, all represented in miniaturized, allegorical form as obstacles on the beach. The games become grander, first involving a man and his child; then a man, his wife, and child; then maybe twenty people; then finally about five hundred players. The last game is cataclysmic; Cribb dies just as Cissie reaches the age of 18.[50]

The project continued to grow in Greenaway's mind. In 1989 he wrote that the plan for the *Fear of Drowning* serial now encompassed nine parts: 'Each episode would increase in length, starting at twenty minutes and increasing in five-minute increments until 115-minute *Drowning by Numbers* was reached.'[51]

POLYGRAPHS

Number charts and graphs are images that are generated from numbers. Increasingly, these images furnish imaginary landscapes that allow us an increasingly intimate and quasi-corporeal inhabitation of a world, not so much of numerical quantities, but numerical relations. Thus we may feel ourselves to be on a 'steep learning curve', or indeed be 'behind the curve', to have 'plateaued out', or be 'in a trough'. We may also be 'on the spectrum', 'at our peak' or sometimes even 'off the scale'. What has become known as the 'uncanny valley' suggests a particular place, but is in fact the region of a graph, described in 1970 by Masahiro Mori, in which the otherwise steadily increasing pleasure in the verisimilitude of an artificial human figure suddenly, at the point of near-indistinguishability, dips into uneasiness.[52] We are nagged incessantly about how bad what are called 'linear' processes are. We worry, or sometimes tell ourselves to take no account of 'outliers'. When we speak metaphorically of getting a larger or smaller slice of the pie, it is no longer clear if we are thinking of an actual pie or the image of a pie, nor are we sure whether 'spikes', in viewing figures or electricity supply, are to be thought of as direct images, of an object or sensation, or secondary images, of the form of which we may be reminded by the shape of a graph. Share prices 'nosedive' or enter 'a spiral', anxieties 'escalate' and profits, like vital signs, can 'flatline'. Audrey Jaffe's *The Affective Life of the Average Man* has mapped the way in which nineteenth-century fiction itself began to show the correlation of individual emotional states with statistical representations of well-being like the share-price graph.[53] During the 1870s, the word 'polygraph' began to be applied to the 'signatures' of various bodily processes – pulse, respiration, blood pressure, skin conductivity. The lie-detecting polygraph was developed for forensic use by John Augustus Larson in 1921, though names like the 'emotograph' and 'respondograph' were tried out by rivals.[54] Larson himself proposed to call his machine the 'cardio-pneumo-psychogram'.[55] One of the most mysterious operations made possible through such bodily mediations is biofeedback, in which subjects prove able to modify bodily functions through focusing on their graphical outputs, through instruments such as the electromyograph, which measures muscle tensions, the thermistor, which measures skin temperature, the electrodermograph, which measures skin conductance, and the electroencephelograph, which measures the amplitudes of electrical activity in the brain. All of these instruments provide graphical displays of variable numerical values, whether in digits or in images.

These graphical images somehow seem to embody the actuality sufficiently concretely for them to allow for immediate modification of the values they encode.

Mathematical visualizations allow for a kind of mathematics that, though generated from and continually transformed by numerical operations, goes far beyond number, or perhaps takes number far beyond itself. All of this depends upon the computer, which reduces things to numerical values in order to transcend number, attaining to the innumerable by passing through the enumerable.

In its capacity to create visualizations, the computer is a self-transcending machine. It creates formalized models that are complex enough to approach the condition of what is being modelled. Richard S. Palais describes the process of using mathematical visualizations to create images of what he calls 'mathematical objects' and processes that show such objects under transformation.[56] When such objects come close to representing not just abstract or idealized models of mathematical functions, but the contingent shapes that approach, or fall away from, the idealized or regular form, we may say that the usual direction of things has been reversed: that the rational has not been derived from an abstraction of the real, but the real has been derived from a hyper-abstraction of the rational. To speak of seeing number now does not mean seeing the ways in which visual appearances approach or diverge from ideal or abstract forms, but rather seeing the numerality of visual forms as such.

This brings about a further, increasingly wonderful inversion. Whereas mathematical modelling has tended to represent numerical relations as waves, trees, streams, spirals and other natural forms, now the natural forms themselves begin to be apprehensible as the pure appearances of number. Under such conditions of visualization a tree becomes readable as a massively detailed mathematical modelling *of itself*, in many dimensions. The visual appearance of the tree is a kind of metaphor for the multi-parameter calculus of different kinds of ratio and quantity – light levels, nutrients, absorption and transpiration of water, exposure to the chances of disease and damage and possibilities of self-defence and repair – that is what the system of physical relations we know as a tree consists of. Among the most striking rhymes between the abstract forms in which information is visually encoded and embodied forms of natural process is the network. Manuel Lima has suggested that the network may be recursively at work in all understanding of information, such that networks are read by other networks: 'It has been proven that networks are a ubiquitous topology in nature, and a type of encoding similar to fractal encoding might exist in our minds . . . Perhaps we have a propensity for structures similar to our own brain.'[57] It is a three-dimensional equivalent of the map drawn on a scale of one mile to the mile that is imagined in Lewis Carroll's *Sylvie and Bruno Concluded*. When Borges reimagined this idea in his short story 'On Exactitude in Science', he added to it a note of melancholy ridicule, since a map on such a scale is seen to be useless, so that it is abandoned and survives only in tattered ruins in desert regions.[58] But Carroll anticipates a much more dramatic step:

> 'It has never been spread out, yet,' said Mein Herr: 'the farmers objected: they said it would cover the whole country, and shut out the sunlight! So we now use the country itself, as its own map, and I assure you it does nearly as well.'[59]

Carroll asks us to imagine a world of forms that might be regarded as their own maps. One of the most amazing embodiments of this is the fact that, viewed from high in the atmosphere, weather systems look so much like the diagrams with which we represent them. Maps and graphs are annotated pictures and, though it is often convenient for all the annotations to be visible and present in the text, digital encoding means that it does not have to be. What has been called 'augmented reality' allows for the information, quantitative and otherwise, to be brought into conjunction with the image when it is needed, in just the way that a mobile navigation system calls up whatever local map may be required at a particular moment. It is this which has persuaded Michel Serres that we have moved from what he calls the declarative to the procedural.[60] A model or a map reduces the complexity of what it models or maps because it aims to show it all at once, in a conspectus, rather than in bit-by-bit detail. This declarative order of a model or map is apparent rather than emergent. Thus, the local fluctuations of suicide rates or numbers of undelivered letters in Paris yield a stable image of a normal distribution that spatializes and thus makes visible a process that, as it emerges in time, is concealed. But digital storage allows for a different form of map, or model of the model, in which one does not need to have the whole map accessible in order to be able to access its relevant portion. The speed of the computer allows for all the information of the map to be kept available in distributed form, which allows one to select only that portion of the map that is needed. I do not need a table on which to unfold the map of Britain, or Australia, in order to locate myself on a map of Paddington.

Thus the whole map, of the country, the genome, or whatever system of information is in question, is virtual and emergent, and its visibility is contingent, modular and on-demand. This makes it more temporal than spatial, especially since the speed with which individual images can be refreshed allows for the display of movements as well as states of data. In this, the map, or the model, come to converge with the nature of the system in itself. A system like an oak tree or an occluded front is not an all-at-once phenomenon: its values, along many different axes and dimensions, are constantly being adjusted. Its visible condition is simply a state of the system. But the fact that this is a system that in principle is capable of being modelled, and in this newly procedural way, allows us to see its contingency, not as the real that escapes or underlies the rational real, the unnumbered form of quantitative relations, but as the projection or diagram-in-the-making of those numerous numerical values. We have turned quantity into geometry or graphic form; and, as a consequence and accompaniment, we have made graphic form full of number. The anomalous name for this amalgam between the graphic and the numerical, anomalous because there is nothing 'given' about it, is 'data'. We see the shape and shaping of number everywhere, not only as the ways in which we might render it quantitatively but as the way in which it shapes itself. Number comes in and out of visibility; and visibility just is this coming in and out of appearance of number. Number is not only mediated in visual form, it mediates between word and image. In our world, the eye is becoming ever more numerate.

One of the most important enabling features of this numerizing of shape and surface is the development of new forms of mathematical geometry, which themselves seem to operate in some realm between appearance and number. Early twentieth-century art responded energetically to the higher-dimensional geometry developed in the second half of the nineteenth century, and a figure such as the knot, which features in the art and magical practice of many cultures, has been made into a newly mathematical figure by the branch of topology known as knot theory. Though much of knot theory appears independent of number, it continues to depend upon numerical relationships. This opens the possibility of discerning the spatialized mathematics at work in the symbolism of different cultures, as suggested in Susanne Küchler's discussion of the function of the knot in making visible complex logical relationships within a culture:

> a knot is not referential but synthetic, in relating inextricably the texture of its surface to the logic of binding. Unlike the open mesh of the looped string, the knot does not hint at what lies beneath its surface, but is itself to be discovered beneath its own surface. The knot is all that is to be seen. The knot is the knowledge, a knowledge of the linking of things, material and mental, that may as well exist apart.[61]

This is very different from the manner of seeing that reduces appearances to the ultimate or essential mathematical forms and relations of which they are the approximations – the Golden Section, or the Platonic Solids. Numbers here are no longer what lie beneath or behind the fluctuations of form, nor do they prescribe the ultimate regularities towards which they tend. Numbers measure the process whereby things move in and out of the condition of form, which is to say of repeatability. Number is therefore immanently at work at every moment. The idea that the world is governed by certain ideal proportions must abstract those proportions from the actual visible world. Numbering used to belong or conduce wholly to the rational, which is to say the regular. In our world, number has gone native, blending with and binding to the irregular, the variant, the stochastic. The visual form of the quantitative imagination used to be the 'geometric' – rational grids, orderly networks and the standardized, soul-sapping 'little boxes' lamented in the Malvina Reynolds song released by Pete Seeger in 1963 – but it is rapidly becoming visible as trees, helices, waves, vortices and radiations of all kinds. Mathematics depends, not just on the symbols that signify quantities, but on the forms of notation developed to perform mathematical operations with those quantities. The forms of nature, drawn into forms of visual quantality, are themselves becoming legible as notations as well as values, not just the result of mathematics, but the animate appearances of mathematics being done. We do mathematics, not just on the world, but in and with it. By a century ago, physics, the study of the visible, tangible and material world, had become almost entirely mathematical. It is less often observed that an inverse process has also been occurring over that century. Precisely because physics has become mathematics, it has become possible for mathematics to become physics, becoming ever more visibly embodied. Turning stuff into number has made it possible for number to become stuff. *Numerus caro factus est*; the number has become flesh.

In one sense, this is a subjection of the world of nature to human processes. The age of the world picture spoken of by Heidegger, in which nature is made available for technological manipulation, has moved into a new phase, in which the physical world is made over into information.[62] As Orit Halperin observes, this means that the process of what is called 'visualization' changes from a psychological to a technical process, undertaken through various kinds of procedure independently of specific acts of human cognition.[63] In such a world, it is not so much that data and calculative procedures are made visible in space, as that 'space becomes "smart" through new models of sense, measure, and calculation' (*Beautiful Data*, 29). A world that is not only the object of calculating rationality but is itself an active participant in that rationality is enabled and accelerated by turning 'a world of ontology, description and materiality to one of communication, prediction, and virtuality' (*Beautiful Data*, 40). Understanding the world as a network of entities in communication with each other required just the generalization of the understanding of communication that was provided by Claude Shannon's mathematical theory of communication.[64] Beginning as an attempt to render the act of communication in mathematical terms, this became a theory of the mathematical basis and functioning of all communication, conceived statistically as information. A general theory of communication, and a theory of general communication, requires the idea of general computation. Communication needs computation perhaps because, as Shannon proposes, communication is computation. It is not just that there are more and more ways of envisaging quantities, magnitudes and numerical relations, nor even that visual space has increasingly been subject to calculative procedures, not least, of course, in the pursuit of various forms of profit, or to avoid various kinds of loss, or minus quantity, which in our world takes on ever more positive and substantially visible forms. It is also that the entry of number and computation into appearance has made possible the autonomization of visual space that Halperin calls 'communicative objectivity', aiming at 'informational infinitude' (*Beautiful Data*, 84) and characterized by 'the production of algorithms, methods, and processes that facilitate interaction, based on the assumption of stored information always/already available' (*Beautiful Data*, 94). In such an epistemological regimen, perception and cognition are no longer distinguished, 'to see and to think being analogized into a single channel' (*Beautiful Data*, 95).

Of course, the appearances of things in the world are not in themselves numerical, for they will always require translation and mediation to turn them into pure number. Whether we are talking about the icon that announces the successful completion of a cash withdrawal, the swinging open of a subway barrier on production of a smartcard or the sequencing of a traffic light, we may say that number, frequency and quantity have entered deeply into appearance. In a sense, these visual signs may be thought of as algebraic: that is, as nonspecific placeholders for numerical values that have no need to be supplied.

BEAUTY

It is clear that there is very great utility in the visual display of quantitative information, especially in making larger patterns evident that might otherwise be lost in the welter of numbers. But there seems to be another gain, which goes beyond utility, in that such displays seem often to be thought, and said to be, beautiful. The importance of beauty in mathematics is one of the most dependable clichés: no conversation about the difference between science and art can proceed for very long without somebody issuing the that's-that announcement: 'But of course mathematicians think beauty is important too.' Many more questions might be asked about this intended showstopper than usually are. Given how much agony there has been among philosophers and critics about precisely what constitutes beauty and how it may be determined, it is charming but odd that mathematicians, or those who speak on their behalf, apparently have no difficulty in knowing it when they see it. The assumption that value in the arts is centred on beauty is also slightly embarrassing; for some considerable time now, critics have been much more interested in art that is interesting rather than beautiful, probably because when it comes to making work interesting they are such interested parties.

In any case, the proliferation of 'beautiful data', or the idea that the visualization of data can produce beauty, provides an occasion to reflect on the work that number might be doing in ideas about beauty and that even more elusive notion, the 'aesthetic'. In order to do this, we would do well to try to substitute some sensible notions for the mystical and overheated discourse to which discussions of beauty can lead. There are, of course, many different kinds of thing that may be seen as beautiful, or visually pleasurable, but amid this diversity there is arguably a large class of things in which the apprehension of pattern, arrangement or symmetry is to the fore in the judgement. It has been suggested that many organisms on earth tend to find symmetrical forms beautiful because symmetry is expensive in terms of natural resources:

> The flower or animal with symmetry is sending out a very clear signal of its genetic superiority over its neighbours. That is why the animal world is populated by shapes that strive for perfect balance. Humans and animals are genetically programmed to look upon these shapes as beautiful – we are attracted to those animals whose genetic make-up is so superior they can use energy to make symmetry.[65]

Though we can be forgiven for not noticing it amid the incessantly designed forms of our human world, orderliness is in fact rare in nature. From the viewpoint of information theory, order involves redundancy, or recurrence, which may be given a mathematical definition as a compression or economy of resources. The more orderly a system is, the closer it will come to an abstract description of the system; the more chaotic and less orderly it is, the less susceptible it is of description by an abstract formula.

This principle was itself formalized neatly by the American mathematician G. D. Birkhoff in his largely forgotten book *Aesthetic Measure* (1933): 'it is the intuitive estimate of the amount of order *O* inherent in the aesthetic object, as compared with its complexity *C*, from which arises the derivative feeling of the aesthetic measure *M* of the different objects of the class involved' – or $M = O/C$.[66] Birkhoff's work was carried forward in the light of the statistical principles of information theory in the work of Max Bense.[67] Birkhoff's title provides a nice and perhaps even recursive ambivalence. On the one hand, it points to a way of measuring the feeling of beauty or satisfaction in works of art – a 'quantitative index of their comparative aesthetic effectiveness'.[68] On the other, it points to the implicit mensuration involved in any feeling of the aesthetic: the pleasure one takes from a work of art involves one in taking the measure of that pleasure. Such a view may seem a little perverse and indigestibly abstract in its form, but it is quite easy to draw it into accord with other, less obviously mathematical but still economic or quantitative, readings of the effects of art, especially psychological theories like those of Freud or I. A. Richards (neither of whom Birkhoff mentions). It might even be squared with Kantian aesthetics, though Birkhoff remarks of Kant's writing only that 'there is evident a strong tendency to adopt a mystical view towards art . . . [and] little that can be regarded as analytical'.[69] Kant distinguishes between the beautiful and the sublime on the grounds that the feeling of the beautiful is 'coupled with the representation of *quality*', whereas that of the sublime is linked to 'that of *quantity*'.[70] This is because, whereas the sensation of the beautiful is related to some, usually natural, object, that of the sublime is related to a judgement or sensation in the observer:

> The beautiful in nature is a question of the form of the object, and this consists in limitation, whereas the sublime is to be found in an object even devoid of form, so far as it immediately involves, or else by its presence provokes, a representation of *limitlessness*, yet with a super-added thought of its totality.[71]

What follows, in sections 25–7 of the *Critique of Judgement*, is a discussion of the mathematical form of the sublime – though, unlike Burke, who takes magnitude or the infinite to be just one form of the sublime, Kant seems to see this mathematical component in all instances of the sublime. He aims to show that '*the sublime is that, the mere capacity of thinking which evidences a faculty of mind transcending every standard of the senses*,' as opposed to more relative and therefore more strictly mathematical judgements of greater or less.[72] It is very hard to see how this therefore proves the sublime to be essentially quantitative, or, rather, it is hard to see how the feeling of the beautiful, which involves the idea of limit, can itself be separated from any kind of quantitative judgement. We may spare ourselves the toil of trying to square Kantian aesthetics in detail with information theory – while nevertheless observing that a kind of quantality, the sense of scale and relative magnitude, attends all of his strange but strangely compelling thinking about so-called aesthetic feelings.

In principle, we might perhaps say that infographics and visualized data may seem beautiful because they allow and display this compression of contingent detail into pattern. But the important thing to note is that economization through pattern does not intensify the sense of beauty in order evenly and uninterruptedly, for there will always come a point at which redundancy will seem merely repetitious and therefore mechanical. Beauty of this kind seems to need the sense of a pattern emerging from the midst of contingency, regulating but not entirely subsuming or subduing that contingency. So when we are assured that number, in the form of visualized data, is beautiful, it may be that we are recognizing the importance of number in the idea of beauty itself. Perhaps all visual pleasure is in part a pleasure in the commanding exercise of vision itself, and perhaps that pleasure is a function of the economic gain in being able to see more than the eye can ordinarily see, or make sense of in what it sees, through various kinds of summing numerical symbolism. Perhaps, that is to say, our pleasure in the many forms in which the statistical is displayed is itself statistical. If there is pleasure in the look of numbers, perhaps it is because of the numbers game that is always being played out in looking.

10

ENOUGH

I have had occasion in the course of this book to consider some of the ways in which, among many other things, the reading and writing of books may themselves be seen and lived as what I have called quantical. The final stages of writing a book certainly seem to be when the process becomes, if not a completely mathematical affair, then certainly a matter more and more of *ratio*, of weights and measures, lengths and proportions, extensions and extractions. In its final stages, in the midst of which I promise I am literally writing these words, a book, which will have been for some years indeterminate as regards its length and components, starts to become more and more a kind of object you can take the measure of, with the possibility of assuming a substantial, material existence in the world. At this point, word-counting, estimating, balancing and the drawing up of accounts start to become more compelling and comprehending procedures than before. One of the great pleasures of seeing a book in proof is finally seeing how many pages it has amounted to, like the outcome of the guess-the-weight-of-the-cake competitions that may still somewhere be played in summer fairs, or the strange ritual of telling people not only that your new baby is safely arrived but also its birthweight (why?). For me, the best part of writing the kinds of book I get to write comes a little further on still, and is actually dependent on there being page numbers, namely, the compiling of an index. I have never understood how authors can delegate this most delicious of tasks to someone else, since it allows one a final, glorious reprise of the whole enterprise, yet without the need to produce a single extra word (while also enabling one last, sometimes surrealistically astonishing, reshuffle of its contents). I am sure that, following Vladimir Nabokov who, in *Pale Fire*, wrote a novel consisting very largely of endnotes, somebody must have written a text consisting solely of the index to some other, absent text.

Agreeable though it is, in a way, to arrive at the point of writing a conclusion, I have never felt quite confident of knowing what a conclusion is for and consequently how to write one. A conclusion may be said to be in part a kind of reassurance for the reader that enough is indeed enough, that they have not been short-changed, or wearied by unnecessary expatiation. But how much is enough? And how much, in a conclusion, is enough to show that there is enough of a book at its back to earn the right to write the conclusion one is writing? Say what you are going to say, say it, then say what you have said, goes the approved formula. But if you have to say what you have said, if there is really something still to say about what you have said, might it not mean that you have not in fact fully said it yet? If you still need a conclusion, in other words, it may be too early for it.

Yet a conclusion must always in some sense also exhibit its own gratuity, confirming for the benefit of both reader and writer that the book actually needs nothing more. This makes a conclusion very literally an exercise in redundancy, in the way in which I have defined it in Chapter Seven, as the amount of information required to transmit a message minus the amount of information needed for the message itself. This implies that a conclusion can only be the right length if there is just a touch too much of it. It can only succeed in confirming that the book it completes is not only (almost) complete, but also entire, if it persuades the reader that the book was in fact already entire before the conclusion that superfluously declares it to be over and out.

Oddly enough, as I sit here doing that very thing, and at the point where it ought to be clearer than ever before how much of the road remains before me, I find myself taken up in a more open-ended kind of proceeding than at almost any point in writing the other parts of the book. How much should I be shooting for in this conclusion? If the function of a conclusion is to demonstrate, without too much fuss or circumstance, that it has no real function to fulfil at all, it can be an unexpectedly ticklish job to judge just how much unnecessariness is strictly necessary at such a juncture.

The conclusion can only really confirm the integrity of the work that it seals and concludes by standing at a little distance from it. The thing that confirms and, as Jacques Derrida would certainly say, countersigns the work is always a sort of outwork, a little gazebo or porter's lodge built at some distance from the main building – or the extra, rather runtish little roll you make with the leftover pastry, that never has quite enough jam in it.

So much, perhaps a little too much, for quantical reflexivity, and on, if possible, to one or two more substantial topics.

It is generally a bad idea to write a book explaining why something else is a bad idea, but I fear I may sometimes have come close to doing that in the book you have just read, or are wondering whether to bother reading. If I have sometimes been unnecessarily and unconvincingly absolute in my characterization and condemnation of the allergy to number and the anti-quantitative prejudice in what I have called the humanities, a term which may be both too spongy and too spikily specific, I would like to think that it has been in the interest of intimating some new ways of thinking about number and quantity. Some time ago, I abandoned what among many academics is regarded as a sacred duty and sustaining vocation, namely the practice of critique, a practice that in its academic forms is usually as pompous as it is footling. But it is not the danger of being thought pompous or footling myself, real and present though that may be, that is my reason for trying to steer clear of critique. The reason is that showing people how bad their ideas are is never a good idea if you genuinely want them to adopt new ones. You may in fact just encourage them to think that if their ideas can only be dismissed in so pompous or footling a fashion, they must have something in them after all. The deployers of critique of course, like the denouncers of sin, are rarely very interested in getting people to think other or better things than the ones that are being critiqued, since that would make further critique unnecessary, a gloomy prospect for those whose salaries and sense of self-worth depend on there being endless things in need of critique. Safer, therefore, to find something as irremediable as it is insufferable against which to rail. Still, if you really are interested in talking people out of bad ideas, the only way is to seduce them with the prospect of new, good ones. Readers of my chapter on the measures of pleasure will know that by good ideas I mean ideas that seem to promise more profit, under whatever understandings of profit will seem to weigh most compellingly, depending. There's no alternative to tasting the cake, or at least putting it on the scales.

Yet I am not completely confident that brand-new ideas are in any case what have been on offer here. Perhaps it is a weakness to which anyone arguing a case of which they are convinced is liable, to be unable to see why anybody could ever have believed anything else. But I have been guided in the arguments I have tried to put together in these chapters, about the importance of number in thinking about death, horror, reading, play, religion, chance, history, jokes, pleasure, sex, music and visual information, by the quiet conviction that we are all of us already virtuosi, if more unconsciously than we might be, in the arts of quantical thinking and feeling. We may not always know if something is worth it, or even how to make the calculation, but I cannot conceive the state of mind of anyone who could not understand the force of such a question. This may be because we necessarily have to ask and answer such questions in everything that we do, as a function of the kinds of being we are – or because, in fact, our individual and collective being cannot be better understood than as itself an ongoing calculus, balancing different kinds of input and output, investment and return, energy and information budgets. Accordingly, I have tried to make it my business less to broach thrilling or bracingly arduous new intellectual possibilities than to articulate some of the kinds of thing that it would be impossible for a sentient being in the twenty-first century not already to know and feel, with respect to what I have called, in a possibly rather numbing shorthand, the order of number. So this is not a book for mathematicians, or not for mathematicians, if there are any, who only want to read about mathematics. My book does not try to explain things we do not understand about number so much as try to explain that, and why, we may not sufficiently realize how much we already do understand, or at least assume about it. We not only do not know how to think and feel outside considerations of more and less, we do not want to know, since we do not know what wanting itself could ever be outside such quantical considerations.

This should not be taken to imply that the force or order of number in human affairs is given and invariant. Not all the arguments made in this book are historical, though, taken as a whole, the book has an historical inflection and itinerary. Ian Hacking's notion that we have been subject since the late nineteenth century to an 'avalanche of number' programmes much of the thinking about modern experiences of quantification. This view implies a kind of historical fall into number, though out of what blessedly Edenic state I cannot myself imagine, for I do not believe that there has ever been a time in which considerations of more and less can ever have been inconspicuous or of no account to human beings. So, if things have changed, the change is itself best understood in quantitative terms, rather than as any kind of qualitative shift. There are, it seems, many more ways than ever before in which questions of more and less may or must be transacted in modern life. For recent and contemporary humans, I have said, number is neither on the side of the human, nor on the side of the inhuman, but in the turbulent middle between them, a middle that number itself animates and agitates. Because number is involved, in many different ways, in these transactions of value, it is dull and dimming to think that one could assign any particular kind of value to number itself. My aim, therefore, has not been to settle the question for or against number, but rather (to employ a bit of committee-speak of which I have always been fond), to move the motion that the question be now put.

Perhaps the most important thing about the case I have made about the awareness and experience of number is that what I just now incautiously called 'the order of number' is not all that orderly and so certainly not uniform. The number of number does not cancel down to one.

Just as little eligible for being counted-as-one is the question of economics. I realize that there may be many nowadays who will assume that to consider the impact of number on modern life is nothing more than to register the reduction of everything in modern life to economics, or, cutting that long story briskly short, the experience of 'capitalism'. A. J. Ayer remarks, of a somewhat earlier commanding idea, that 'a hypothesis which explains everything, in the sense that nothing is counted as refuting it, explains nothing.'[1]

One might press further and say that the more something explains, the less likely it is in fact to exist. And there seems truly to be nothing that the fact of capitalism, the greatest and most indubitable Count-As-One of our time, is not thought to be able to explain. If I have not discussed economics in any straightforward and set-apart way, I hope it will be clear from what I have written that this is not at all because I doubt its importance. On the contrary, it is because I am convinced that economics is at work always and everywhere in human affairs, but in so pervasive and so variegated a way that it makes no kind of sense to suggest that economic questions might be quarantined from all other areas. In this respect, and encouraged by the perspectives I take from information theory, I imply and assume a general economy of economies, monetary, sexual, biological, linguistic, religious and political.

Such a view sits well with my prejudice that the only kind of subject on which it is worthwhile to write a book is one that a single book has no chance of summing up. So I am reassured rather than rattled by the thought of the many different topics that might have been considered in this book but, for almost entirely quantical reasons, have not been. (Where are the discussions of calendars, or crosswords, or the chapter I meant to write on the imagination of negatives and minus quantities, for example? Brooding yet, if anywhere, in the teeming womb of time.) You can only finish a book, people are weary of me telling them, if you are able to imagine a plausible string of sequels to it. A large part of the pleasure that accompanies the cooling of the *furor scribendi* in finishing a book of this kind is that it frees one to imagine completely different ways in which the whole enterprise might have been undertaken. You may recognize here the well-known device of using a conclusion to demonstrate that, far from being any kind of definitive statement, the preceding book has in fact been no more than a prolegomenon to the real work that remains to be done, like the plumber who uses the opportunity of replacing a tap washer to demonstrate that you need an entire new central heating system. In any case, I hope not to have had the final word or got anywhere near the bottom line as far as quantality and the culture of number are concerned. That will have to be, at least for me, for now, enough.

REFERENCES

ONE: VERNACULAR MATHEMATICS

1 George Eliot, *Adam Bede*, ed. Carol A. Martin (Oxford, 2008), p. 209.
2 Samuel Beckett, *Collected Shorter Prose, 1945–1980* (London, 1984), p. 80.
3 Terence (Publius Terentius Afer), *The Woman of Andros. The Self-tormentor. The Eunuch*, ed. and trans. John Barsby (Cambridge, MA, 2001), pp. 185–7.
4 Ian Dury, 'Police Will Not Probe Break-ins at Homes With Odd Number', *Daily Mail* (6 August 2015), pp. 1–2.
5 Ian Hacking, *The Taming of Chance* (Cambridge, 1990), p. 2.
6 Ibid., p. 41.
7 Ibid., p. 141.
8 Ibid., p. 63.
9 William Petty, *Five Essays in Political Arithmetick* (London, 1687).
10 Hacking, *Taming of Chance*, p. 141.
11 Quentin Meillassoux, *The Number and the Siren: A Decipherment of Mallarmé's Coup de Dés*, trans. Robin Mackay (New York, 2012), p. 39.
12 Ibid., pp. 45–6.
13 Ibid., p. 74.
14 Margaret Cavendish, *The Blazing World, and Other Writings*, ed. Kate Lilley (London, 1994), p. 171.
15 Ibid., p. 172.
16 Ibid.
17 George Lakoff and Rafael Nuñez, *Where Mathematics Comes From: How the Embodied Mind Brings Mathematics Into Being* (New York, 2001), p. xi.

TWO: QUANTALITY

1 Alain Badiou, *Number and Numbers*, trans. Robin Mackay (Cambridge and Malden, MA, 2008), pp. 89–92.

2 Jorge Luis Borges, *Collected Fictions*, trans. Andrew Hurley (New York and London, 1998), p. 229.
3 Nathaniel Fairfax, *A Treatise of the Bulk and Selvedge of the World: Wherein the Greatness, Littleness, and Lastingness of Bodies are Freely Handled* (London, 1674), p. 110.
4 James Joyce, *Finnegans Wake*, 4th edn (London, 1975), p. 213.
5 A. N. Whitehead, *Religion in the Making* (Cambridge, 2011), p. 5.
6 William Shakespeare, *Othello*, 3rd Arden edn, ed. E.A.J. Honigmann (Walton-on-Thames, 1996), Act I Scene 3, p. 159.
7 A. N. Whitehead, *The Concept of Nature* (Cambridge, 1920), pp. 142–3.
8 Oliver Sacks, *A Leg to Stand On* (London, 2012), pp. 112–13.
9 Ibid., p. 113.
10 Ibid.
11 Ibid., p. 121.
12 Michel Serres, *L'Incandescent* (Paris, 2003), pp. 369–70.
13 Lucretius (Titus Lucretius Carus), *On the Nature of the Universe*, trans. R. E. Latham (London, 1994), 2.402–8, p. 47.
14 Whitehead, *Religion in the Making*, p. 112.
15 Ibid.
16 John Allin and Thomas Shepard, *A Defence of the Answer Made Unto the Nine Questions or Positions Sent from New-England, Against the Reply Thereto by that Reverend Servant of Christ, Mr. John Ball . . .* (London, 1648), p. 46.
17 William Shakespeare, *Love's Labour's Lost*, 3rd Arden edn, ed. Henry Woodhuysen (London, 1998), Act V Scene 2, p. 264.
18 James Elphinstone, *Propriety Ascertained in Her Picture; or, Inglish Speech and Spelling Rendered Mutual Guides*, 2 vols (London, 1786–7), vol. II, pp. 83, 109.
19 Francis Thompson, *Literary Criticisms of Francis Thompson: Newly Discovered and Collected*, ed. Terence L. Connolly (New York, 1948), p. 303.
20 Ibid., p. 302.
21 Ibid., pp. 304, 303.
22 Ibid., p. 303.
23 Lisa Heschong, *Thermal Delight in Architecture* (Cambridge, MA, 1979).
24 Sigmund Freud, 'The Antithetical Meaning of Primal Words', *The Standard Edition of the Complete Psychological Works of Sigmund Freud*, trans. James Strachey, vol. XI (London, 1957), pp. 155–61.
25 William Shakespeare, *As You Like It*, 3rd Arden edn, ed. Juliet Dusinberre (London, 2006), Act III Scene 5, p. 285.
26 Samuel Beckett, *Molloy. Malone Dies. The Unnamable* (London, 1973), p. 179; Samuel Beckett, *Malone meurt* (Paris, 1971), p. 7.
27 Garrett Stewart, *Death Sentences: Styles of Dying in British Fiction* (Cambridge, MA, 1984), p. 321.

28 William Shakespeare, *Sonnets*, 3rd Arden edn, ed. Katherine Duncan-Jones (Walton-on-Thames, 1997), 367.
29 John Keats, *Poetical Works*, ed. H. W. Garrod (London, Oxford and New York, 1970), p. 208; Samuel Beckett, *More Pricks Than Kicks*, ed. Cassandra Nelson (London, 2010), p. 14.
30 Emily Dickinson, *Complete Poems*, ed. Thomas H. Johnson (London, 1970), p. 124.
31 Steven Connor, 'Spelling Things Out', *New Literary History*, XLV (2014), pp. 183–97.
32 Steven Connor, 'Michel Serres: The Hard and the Soft' (2009). Online at http://stevenconnor.com/hardsoft.html.
33 James Joyce, *Ulysses*, ed. Jeri Johnson (Oxford, 1993), p. 46.
34 Peter Sloterdijk, *Spheres*, vol. I: *Bubbles: Microspherology*, trans. Wieland Hoban (Los Angeles, CA, 2011), p. 12.
35 Ibid., p. 51.
36 Ibid., p. 24.
37 Jeremy Gray, *Plato's Ghost: The Modernist Transformation of Mathematics* (Princeton, NJ, 2009).
38 A. N. Whitehead, *Science and the Modern World* (New York, 1948), p. 34.
39 Badiou, *Number and Numbers*, pp. 1, 3.
40 Christopher Small, *Musicking: The Meanings of Performing and Listening* (Hanover, NH, 1998).
41 Michel Serres, *Rameaux* (Paris, 2007), p. 184 (my translation).
42 Ibid., p. 185.
43 William James, *The Principles of Psychology*, 2 vols (New York, 1890), vol. I, pp. 318–19.
44 Claude Shannon, 'A Mathematical Theory of Communication', *Bell System Technical Journal*, XXVII (1948), pp. 379–423, 623–56. Online at http://worrydream.com.
45 Michel Serres, *Hermes: Literature, Science, Philosophy*, trans. Josué V. Harari and David F. Bell (Baltimore, MD, and London, 1982), p. 79.
46 Ibid., p. 80.
47 Ibid., p. 81.
48 Ibid.

THREE: HORROR OF NUMBER

1 Richard Rorty, *Philosophy and Social Hope* (London, 1999), p. 52.
2 Ibid.
3 Ibid., p. 53.
4 G. H. Hardy, *A Mathematician's Apology* (Cambridge, 1992), p. 37.
5 Lewis Carroll, *Alice's Adventures in Wonderland and Through the Looking-Glass: And What Alice Found There*, ed. Peter Hunt (Oxford, 2009), p. 226.

6 Stanislas Dehaene, *The Number Sense: How the Mind Creates Mathematics* (London, 1999), p. 76.
7 Frank Kermode, *The Sense of an Ending: Studies in the Theory of Fiction* (Oxford, 2000), pp. 45–6.
8 Jean-François Lyotard, *Libidinal Economy*, trans. Iain Hamilton Grant (Bloomington and Indianapolis, IN, 1993), p. 164.
9 Kermode, *Sense of an Ending*, p. 46.
10 Marquis de Sade, *The 120 Days of Sodom and Other Writings*, ed. and trans. Austryn Wainhouse and Richard Seaver (New York, 1966).
11 Samuel Beckett, *No's Knife: Collected Shorter Prose, 1945–66* (London, 1967), p. 9.
12 Elizabeth Sewell, *The Field of Nonsense* (London, 1952), p. 65.
13 Ibid.
14 Ibid., p. 67, my emphasis.
15 James Boswell, *Life of Johnson*, ed. R. W. Chapman (Oxford, 2008), p. 339.
16 Bertrand Russell, *Introduction to Mathematical Philosophy* (London, 1920), p. 56.
17 Sigmund Freud, 'The Medusa's Head', *The Standard Edition of the Psychological Works of Sigmund Freud*, vol. XVIII, trans. James Strachey (London, 1955), p. 273.
18 Carroll, *Alice*, p. 105.
19 Norman O. Brown, *Love's Body* (Berkeley and Los Angeles, CA, 1966), p. 66 (quoting Géza Róheim).
20 Noel Carroll, *The Philosophy of Horror: or Paradoxes of the Heart* (New York, 1990), pp. 32–3.
21 Abraham Seidenberg, 'The Ritual Origin of Counting', *Archive for the History of Exact Sciences*, II (1962), p. 9.
22 E. T. Bell, *The Magic of Numbers* (New York, 1946), p. 161.
23 William Shakespeare, *The Merry Wives of Windsor*, 3rd Arden edn, ed. Giorgio Melchiori (London, 2000), Act V Scene 1, p. 271; *Antony and Cleopatra*, 3rd Arden edn, ed. John Wilders (London, 1995), Act IV Scene 15, p. 269.
24 Sigmund Freud, *Beyond the Pleasure Principle*, in *The Standard Edition of the Psychological Works of Sigmund Freud*, vol. XVIII, trans. James Strachey (London, 1955), p. 54.
25 Samuel Beckett, *Worstward Ho* (London, 1983), pp. 32, 33.

FOUR: MODERN MEASURES

1 Charlotte Brontë, *Jane Eyre*, ed. Stevie Davies (London, 2006), p. 82.
2 George Eliot, *Middlemarch*, ed. Rosemary Ashton (London, 1994), p. 140.

3 Ralph Waldo Emerson, *Nature and Selected Essays*, ed. Larzer Ziff (London and New York, 2003), p. 335.
4 Ibid., pp. 335–6.
5 Ibid., p. 223.
6 Ibid., p. 252.
7 Ibid., p. 295.
8 Walter Pater, *Studies in the History of the Renaissance*, ed. Matthew Beaumont (Oxford, 2010), p. 119.
9 Ibid., p. 120.
10 Matthew Arnold, *Culture and Anarchy*, ed. Jane Garnett (Oxford, 2006), p. 123.
11 Ibid., p. 81.
12 F. R. Leavis, 'Mass Civilization and Minority Culture', in *Popular Culture: A Reader*, ed. Raiford A. Guins and Omayra Zaragoza Cruz (London, 2005), p. 36.
13 Ibid., p. 33.
14 Francis Galton, 'The Measure of Fidget', *Nature* (1885), XXXII, pp. 174–5.
15 Francis Galton, 'Arithmetic by Smell', *Psychological Review*, I (1892), pp. 61–2.
16 Edmund Husserl, *Philosophy of Arithmetic: Psychological and Logical Investigations; With Supplementary Texts from 1887–1901*, trans. Dallas Willard (Dordrecht and London, 2003).
17 Mary Poovey, *A History of the Modern Fact: Problems of Knowledge in the Sciences of Wealth and Society* (Chicago, IL, 1998), p. 29.
18 W. H. Auden, *Collected Shorter Poems, 1927–1957* (London, 1969), pp. 319–20.
19 Ibid., p. 320.
20 E. M. Forster, *Howards End* (New York, 1989), p. 30.
21 John Carey, *The Intellectuals and the Masses: Pride and Prejudice Among the Literary Intelligentsia* (London, 1992).
22 D. H. Lawrence, *Complete Poems*, ed. Vivian de Sola Pinto and F. Warren Roberts (Harmondsworth, 1977), p. 431.
23 Ibid.
24 Ibid., p. 441.
25 Ibid., pp. 355–6.
26 D. H. Lawrence, *Women in Love* (Harmondsworth, 1960), p. 115.
27 Ibid., p. 525.
28 Ibid., pp. 117–18.
29 D. H. Lawrence, *Phoenix II: Uncollected, Unpublished, and Other Prose Works*, ed. Warren Roberts and Harry T. Moore (New York, 1970), p. 266.
30 Ibid.

31 Virginia Woolf, *To the Lighthouse*, ed. Hermione Lee (London, 2000), p. 214.
32 Lewis Mumford, *Technics and Civilization* (London, 1934), p. 232.
33 Virginia Woolf, *Selected Essays*, ed. David Bradshaw (Oxford, 2010), p. 9.
34 Kim Shirkhani, 'Small Language and Big Men in Virginia Woolf', *Studies in the Novel*, XLIII (2011), p. 56.
35 Woolf, *Selected Essays*, p. 9.
36 Virginia Woolf, *Mrs Dalloway*, ed. G. Patton Wright (London, 1992), p. 14.
37 Virginia Woolf, *The Years*, ed. Steven Connor and Susan Hill (London, 2000), p. 254.
38 Henri Bergson, *Time and Free Will: An Essay on the Immediate Data of Consciousness*, trans. F. L. Pogson (London, 1910), p. 227.
39 James Joyce, *Letters*, 3 vols, ed. Stuart Gilbert and Richard Ellmann (London, 1957–66), vol. II, p. 134.
40 May Sinclair, *Life and Death of Harriett Frean* (London, 1980), p. 1. References, to HF, in the text hereafter.
41 Samuel Beckett, *Complete Dramatic Works* (London, 1984), p. 476.
42 Michel Serres, *Récits d'humanisme* (Paris, 2006), p. 41.
43 Walter Benjamin, *Illuminations: Essays and Reflections*, trans. Harry Zohn, ed. Hannah Arendt (New York, 1969), p. 236.
44 James Joyce, *Ulysses*, ed. Jeri Johnson (Oxford, 1993), p. 267.
45 James Joyce, *Finnegans Wake* (London, 1971), p. 8.
46 Joyce, *Ulysses*, pp. 279, 267.
47 Garrett Stewart, 'Cinécriture: Modernism's Flicker Effect', *New Literary History*, XXIX (1998), p. 729.
48 Anthony Burgess, *Re Joyce* (New York, 1969), p. 69.
49 Quoted in Oliver Lodge, *The Ether of Space* (New York and London, 1909), p. 113.
50 Joyce, *Ulysses*, p. 220.
51 Hermann Helmholtz, *On the Sensations of Tone as a Physiological Basis for the Theory of Music*, trans. Alexander Ellis (London, 1875).
52 Ibid., p. 46.
53 Bertrand Russell, 'The Philosophy of Bergson', *Monist*, XXII (1912), p. 326.
54 Henri Bergson, *The Meaning of the War: Life and Matter in Conflict* (London and Edinburgh, 1915), pp. 46–7.
55 Henri Bergson, *Matter and Memory*, trans. Nancy Margaret Paul and W. Scott Palmer (London, 1911), p. 268.
56 Ibid.
57 Ibid.
58 Ibid., pp. 265, 266.

59 Ibid., p. 260.
60 Bergson, *Time and Free Will*, p. 12.
61 Henri Bergson, *Creative Evolution*, trans. Arthur Mitchell (New York, 1911), pp. 270–71.
62 Russell. 'The Philosophy of Bergson', p. 333.
63 Eric A. Cornell and Carl E. Wieman, 'The Bose–Einstein Condensate', *Scientific American*, CCLXXVIII (1998), pp. 40–45.
64 Gerard Manley Hopkins, *Poems*, ed. W. H. Gardner and N. H. Mackenzie, 4th edn (London, 1970), p. 52.
65 Lynda Nead, *The Haunted Gallery: Painting, Photography, Film, c. 1900* (New Haven, CT, and London, 2007), p. 194.
66 Stewart, 'Cinécriture: Modernism's Flicker Effect', p. 731.
67 Joseph Conrad, *The Secret Agent: A Simple Tale* (Harmondsworth, 1980), p. 127.
68 Ibid., p. 131.
69 Virginia Woolf, *A Writer's Diary: Being Extracts From the Diary of Virginia Woolf*, ed. Leonard Woolf (London, 1981), p. 17. References, to WD, in the text hereafter.
70 Virginia Woolf, *The Waves* (London, 1980), p. 170.
71 Ibid., p. 102.
72 Ibid., p. 20.
73 Ibid., p. 27.
74 Woolf, *The Years*, p. 254.
75 Samuel Beckett, *Molloy. Malone Dies. The Unnamable* (London, 1973), p. 18.
76 Ibid.
77 Bergson, *Matter and Memory*, p. 260.
78 Christopher Caudwell, *Illusion and Reality: A Study of the Sources of Poetry* (London, 1950), p. 3.

FIVE: LOTS

1 Samuel Beckett, *Complete Dramatic Works* (London, 1984), p. 116.
2 W. B. Yeats, *Essays and Introductions* (London, 1961), p. 215.
3 Ibid., p. 216.
4 Ibid.
5 Immanuel Kant, *Critique of Judgement*, trans. James Creed Meredith, revd and ed. Nicholas Walker (Oxford, 2007), p. 82.
6 Ibid., p. 83.
7 Ibid., p. 81.
8 Kant, *Critique of Judgement*, pp. 80–81.
9 I-Hwa Yi, *Korea's Pastimes and Customs: A Social History* (Paramus, NJ, 2006), p. 227.

10 Ben Jonson, *The Cambridge Edition of the Works of Ben Jonson*, ed. David Bevington, Martin Butler and Ian Donaldson, 7 vols (Cambridge, 2012), vol. II, pp. 402–3; Thomas Hobbes, *Leviathan*, ed. Richard Tuck (Cambridge, 1996), p. 82; Richard Levin, 'Counting Sieve Holes in Jonson and Hobbes', *Notes and Queries*, XLIX (2002), p. 250.
11 Pieter W. van der Horst, 'Sortes: Sacred Books as Oracles in Late Antiquity', in *The Use of Sacred Books in the Ancient World*, ed. L. V. Rutgers, P. W. van der Horst, H. W. Havelaar and L. Teugels (Leuven, 1988), p. 149.
12 Joshua Trachtenberg, *Jewish Magic and Superstition: A Study in Folk Religion* (Philadelphia, PA, 2004), p. 216.
13 Van der Horst, 'Sortes', p. 157.
14 Louis Jacobs, *The Jewish Religion: A Companion* (Oxford, 1995), p. 132.
15 Van der Horst, 'Sortes', p. 165.
16 Trachtenberg, *Jewish Magic and Superstition*, p. 209.
17 Michel Foucault, *The Order of Things: An Archaeology of the Human Sciences* (London, 2002), p. 19.
18 Sammy Githuku, 'Taboos on Counting', in *Interpreting the Old Testament in Africa: Papers from the International Symposium on Africa and the Old Testament in Nairobi, October 1999*, ed. Mary Getui, Knut Holter and Victor Zinkuratire (New York, 2001), pp. 113–18.
19 Abraham Seidenberg, 'The Ritual Origin of Counting', *Archive for History of Exact Sciences*, II (1962), p. 15.
20 Claudia Zaslavsky, *Africa Counts: Number and Pattern in African Cultures*, 3rd edn (Chicago, IL, 1999), p. 51.
21 Seidenberg, 'Ritual Origin of Counting', p. 16.
22 Michael Hardt and Antonio Negri, *Multitude: War and Democracy in the Age of Empire* (London, 2004), pp. 99, xiv. References, to *Multitude*, in the text hereafter.
23 Peter Sloterdijk, *God's Zeal: The Battle of the Three Monotheisms*, trans. Wieland Hoban (Cambridge, 2009), p. 96.
24 Chris Harman, *Zombie Capitalism: Global Crisis and the Relevance of Marx* (Chicago, IL, 2010), p. 12.
25 Jacques Derrida, *Specters of Marx: The State of the Debt, the Work of Mourning, and the New International*, trans. Peggy Kamuf (New York, 2006), p. 57.
26 Joerg Rieger and Kwok Pui-lan, *Occupy Religion: Theology of the Multitude* (Plymouth, 2012), pp. 59–61.
27 Bernhard Citron, 'The Multitude in the Synoptic Gospels', *Scottish Journal of Theology*, VII (1954), p. 410.
28 Jinkwan Kwon, 'Minjung (the Multitude), Historical Symbol of Jesus Christ', *Asia Journal of Theology*, XXIV (2010), pp. 153–71; Hyo-Dong Lee, *Spirit, Qi, and the Multitude: A Comparative Theology for the Democracy*

of Creation (New York, 2014), p. 229.
29 Michel Serres, *Genesis*, trans. Geneviève James and James Nielson (Ann Arbor, MI, 1995), pp. 2–3.
30 Ibid., p. 2.
31 Ibid., pp. 4–5.
32 Edward Smedley, Hugh James Rose and Henry John Rose, eds, *Encyclopaedia Metropolitana, or, Universal Dictionary of Knowledge . . .* 29 vols (London, 1845), vol. I, p. 391.
33 Jean-Paul Sartre, *Being and Nothingness: An Essay on Phenomenological Ontology*, trans. Hazel Barnes (London, 1984), p. 351.

SIX: HILARIOUS ARITHMETIC

1 Michel Serres, *The Parasite*, trans. Lawrence R. Schehr (Minneapolis, MN, and London, 2007), p. 86.
2 Brian Rotman, *Ad Infinitum: The Ghost in Turing's Machine: Taking God Out of Mathematics and Putting the Body Back In* (Stanford, CA, 1993).
3 Immanuel Kant, *Critique of Judgement*, trans. James Creed Meredith, revd and ed. Nicholas Walker (Oxford, 2007), p. 161.
4 Ibid.
5 Ibid.
6 Ibid.
7 Ibid., p. 162.
8 Sigmund Freud, *Jokes and Their Relation to the Unconscious, The Standard Edition of the Psychological Works of Sigmund Freud*, vol. VIII, trans. James Strachey (London, 1960), p. 53.
9 Ibid., p. 59.
10 Elizabeth Sewell, *The Field of Nonsense* (London, 1952), p. 65.
11 M.B.W. Tent, *The Prince of Mathematics: Carl Friedrich Gauss* (Wellesley, MA, 2006), pp. 33–5.
12 Samuel Beckett, *Watt* (London, 1972), p. 46.
13 Charles Dickens, *Hard Times*, ed. Paul Schlicke (Oxford, 2008), p. 8.
14 Charles Dickens, *Great Expectations*, ed. Angus Calder (Harmondsworth, 1965), p. 84.
15 Ibid., pp. 84–5.
16 Ibid., p. 85.
17 Ibid., p. 96.
18 Henri Bergson, *Laughter: An Essay on the Meaning of the Comic*, trans. Cloudesley Brereton and Fred Rothwell (New York, 1914), pp. 4–5.
19 Ibid., p. 4.
20 Beckett, *Watt*, p. 171.
21 Ibid., pp. 173–4.
22 Ibid., pp. 178–9.

23 Ludwig Wittgenstein, *Philosophical Remarks*, ed. Rush Rhees, trans. Raymond Hargreaves and Roger White (Oxford, 1975), p. 128.
24 Lewis Carroll, *Alice's Adventures in Wonderland and Through the Looking-Glass: And What Alice Found There*, ed. Peter Hunt (Oxford, 2009), p. 104.
25 Chris Ackerley, *Obscure Locks, Simple Keys: The Annotated 'Watt'* (Tallahassee, FL, 2005), pp. 160–61.
26 G. H. Hardy, *A Mathematician's Apology* (Cambridge, 1992), pp. 150–51.
27 Ibid., p. 151.
28 Charles Dickens, *Martin Chuzzlewit*, ed. Margaret Cardwell (Oxford and New York, 1991), p. 66.
29 Fredric Jameson, *The Political Unconscious: Narrative as Socially Symbolic Act* (Ithaca, NY, 1981), p. 102.
30 Steven Connor, *A Philosophy of Sport* (London, 2011), pp. 151–6.
31 Peter Sloterdijk, *Rage and Time: A Psychopolitical Investigation*, trans. Mario Wenning (New York, 2010).
32 Gottfried Wilhelm von Leibniz, *Leibniz: Selections*, ed. Philip P. Wiener (New York, 1951), p. 51.
33 Jeremy Bentham, *Introduction to the Principles of Morals and Legislation* (Oxford, 1907), p. 2. References, to *Introduction*, in the text hereafter.
34 Wesley C. Mitchell, 'Bentham's Felicific Calculus', *Political Science Quarterly*, XXXIII (1918), p. 180.
35 Philip Larkin, *Collected Poems*, ed. Anthony Thwaite (London, 1990), p. 147.
36 Ibid., pp. 89, 148.
37 Michel Serres, *The Parasite*, trans. Lawrence R. Schehr (Minneapolis, MN, and London, 2007), p. 87.

SEVEN: PLAYING THE NUMBERS

1 Thomas Hardy, *Complete Poems*, ed. James Gibson (Basingstoke, 2001), p. 7.
2 Robert Newsom, *A Likely Story: Probability and Play in Fiction* (New Brunswick, NJ, and London, 1988).
3 Nassim Nicholas Taleb, *The Black Swan: The Impact of the Highly Improbable* (London, 2007), p. 127.
4 F. N. David, *Games, Gods and Gambling* (London, 1962), pp. 7–8.
5 Roland Barthes, 'Réflexions sur un manuel', in *L'enseignement de la littérature*, ed. Serge Doubrovsky and Tzvetan Todorov (Paris, 1981), p. 64.
6 Franco Moretti, *Graphs, Maps, Trees: Abstract Models for Literary History* (London, 2007), p. 92. References, to *Graphs, Maps, Trees*, in the text hereafter.

7 Philip Ball, *Critical Mass: How One Thing Leads to Another* (London, 2004).
8 Leonard Mlodinow, *The Drunkard's Walk: How Randomness Rules Our Lives* (London, 2008), pp. 165–8.
9 Daniel C. Dennett, *Darwin's Dangerous Idea: Evolution and the Meanings of Life* (London, 1996), p. 65.
10 Tristan Tzara, 'Pour faire un poème dadaïste', *Littérature*, XV (1920), p. 18 (my translation).
11 Oscar Wilde, *The Importance of Being Earnest and Other Plays*, ed. Peter Raby (Oxford, 2008), p. 265.
12 Stanley Fish, *Doing What Comes Naturally: Change, Rhetoric, and the Practice of Theory in Literary and Legal Studies* (Oxford, 1989), p. 91.
13 William Shakespeare, *Antony and Cleopatra*, 3rd Arden edn, ed. John Wilders (London and New York, 1995), Act V Scene 2, p. 276.
14 Natasha Lushetich, 'Ludus Populi: The Practice of Nonsense', *Theatre Journal*, LXIII (2011), pp. 33, 34.
15 Ibid., pp. 29, 35.
16 Ibid., p. 34.
17 Gary Saul Morson, 'Contingency and Poetics', *Philosophy and Literature*, XXII (1998), p. 295.
18 Ibid., p. 300.
19 Keith Devlin, *The Unfinished Game: Pascal, Fermat, and the Seventeenth-century Letter That Made the World Modern* (New York, 2008).
20 Deborah J. Bennett, *Randomness* (Cambridge, MA, and London, 1998), pp. 132–51.

EIGHT: KEEPING THE BEAT

1 George Puttenham, *The Art of English Poesie*, ed. Gladys Doidge Willcok and Alice Walker (Cambridge, 1936), p. 77.
2 Oliver Goldsmith, *Works*, 4 vols (Edinburgh, 1892), vol. I, p. 385.
3 Edward Wadham, *English Versification: A Complete Practical Guide to the Whole Subject* (London, 1869), p. 114.
4 Ibid.
5 John Dryden, *Sylvae, or, The Second Part of Poetical Miscellanies* (London, 1684), sig. a6v.
6 John Foster, *An Essay on the Different Nature of Accent and Quantity, With Their Use and Application in the Pronunciation of the English, Latin, and Greek Languages* (Eton, 1762), pp. 98, 99.
7 Samuel Parr, *A Discourse on Education and on the Plans Pursued in Charity-schools* (London, 1785), p. 3.
8 Thomas De Laune and Benjamin Keach, *Tropologia, or, A Key to Open Scripture Metaphors* (London, 1681), p. 3.

9 Thomas Sprat, 'An Account of the Life of Mr. Abraham Cowley', in *The Works of Mr. Abraham Cowley* (London, 1668), sig. b1v.
10 Geoffrey Chaucer, *The Canterbury Tales of Chaucer, In the Original, From the Most Authentic Manuscripts* . . . ed. Thomas Morell (London, 1737), pp. xxv–xxvi.
11 Stanislas Dehaene, *The Number Sense: How the Mind Creates Mathematics* (London, 1999).
12 Simon Jarvis, 'Prosody as Cognition', *Critical Quarterly*, XL (1998), pp. 5, 6.
13 Ibid., p. 10.
14 Simon Jarvis, 'The Music of Thinking: Hegel and the Phenomenology of Prosody', *Paragraph*, XXVIII (2005), p. 58.
15 Ibid., p. 64.
16 Ibid., p. 63.
17 Jean-Luc Nancy, *Listening*, trans. Charlotte Mandell (New York, 2007), p. 17.
18 Ibid., p. 21.
19 A. P. Juschkewitsch and Ju. Ch. Kopelewitsch, 'La correspondance de Leibniz avec Goldbach', *Studia Leibnitiana*, XX (1988), p. 182.
20 Oliver Sacks, *The Man Who Mistook His Wife for a Hat* (London, 2007), p. 215.
21 Arthur Schopenhauer, *The World as Will and Representation*, 2 vols, trans. E.F.J. Payne (New York, 1969), vol. I, p. 256.
22 Gottfried Wilhelm von Leibniz, *Philosophical Papers and Letters: A Selection*, ed. and trans. Leroy E. Loemker, 2nd edn (Dordrecht, Boston and London, 1989), p. 641.
23 Oliver Sacks, *Musicophilia: Tales of Music and the Brain* (London, 2008), p. 255.
24 Ibid., p. 256.
25 Ibid., p. 260 n.2.
26 William Shakespeare, *As You Like It*, 3rd Arden edn, ed. Juliet Dusinberre (London, 2006), Act V Scene 4, p. 344.
27 John Perceval, *A Narrative of the Treatment Experienced by a Gentleman, During a State of Mental Derangement; Designed to Explain the Causes and the Nature of Insanity, and to Expose the Injudicious Conduct Pursued Towards Many Unfortunate Sufferers Under That Calamity* (London, 1840), pp. 304–5.
28 Richard Feynman, *What Do You Care What Other People Think? Further Adventures of a Curious Character* (London, 2007), p. 57.
29 Lewis Carroll, *Alice's Adventures in Wonderland and Through the Looking-Glass: And What Alice Found There*, ed. Peter Hunt (Oxford, 2009), p. 63.
30 R.H.F. Scott, *Jean-Baptiste Lully* (London, 1973), pp. 115–17.
31 Eve Kosofsky Sedgwick, *Tendencies* (Durham, NC, 1993), p. 181.

References, to *Tendencies*, in the text hereafter.

32 Sigmund Freud, 'A Child is Being Beaten', *The Standard Edition of the Psychological Works of Sigmund Freud*, vol. XVII, trans. James Strachey et al. (London, 1955), pp. 179, 185.

33 William Niederland, 'Early Auditory Experiences, Beating Fantasies, and Primal Scene', *Psychoanalytic Study of the Child*, XIII (1958), pp. 471–504.

34 Simon Jarvis, 'The Melodics of Long Poems', *Textual Practice*, XXIV (2010), p. 609.

35 Ibid.

36 Ibid., p. 617.

37 Ibid., p. 618.

38 Michel Serres, *Musique* (Paris, 2011), pp. 43–4 (my translation).

39 Konstantin V. Zenkin, 'On the Religious Foundations of A. F. Losev's Philosophy of Music', *Studies in East European Thought*, LVI (2004), p. 161.

40 Jarvis, 'Prosody as Cognition', p. 6.

41 Fred Kersten, 'Can Sartre Count?', *Philosophy and Phenomenological Research*, XXXIV (1974), p. 342.

42 Schopenhauer, *World as Will and Representation*, vol. I, p. 264; Kersten, 'Can Sartre Count?', p. 353.

NINE: THE NUMERATE EYE

1 Thomas Middleton and William Rowley, *The Changeling*, ed. Tony Bromham (Basingstoke, 1986), Act II Scene 2, p. 69.

2 William Shakespeare, *Hamlet*, 3rd Arden edn, ed. Ann Thompson and Neil Taylor (London, 2006), Act V Scene 2, p. 441.

3 William Shakespeare, *Troilus and Cressida*, 3rd Arden edn, ed. David Bevington (Walton-on-Thames, 1998), Act II Scene 3, pp. 257–8.

4 Katherine Hunt, 'Convenient Characters: Numerical Tables in William Godbid's Printed Books', *Journal of the Northern Renaissance*, VI (2014). Online at www.northernrenaissance.org.

5 E. L. Kaufman, M. W. Lord, T. W. Reese and J. Volkmann, 'The Discrimination of Visual Number', *American Journal of Psychology*, LXII (1949), pp. 498–525.

6 Stanislas Dehaene, *The Number Sense: How the Mind Creates Mathematics* (London, 1999), pp. 13–40.

7 Karl Menninger, *Number Words and Number Symbols: A Cultural History of Numbers*, trans. Paul Broneer (Cambridge, MA, and London, 1969), p. 317.

8 Katherine Hunt and Rebecca Tomlin, 'Editorial: Numbers in Early Modern Writing', *Journal of the Northern Renaissance*, VI (2014). Online

at www.northernrenaissance.org.

9 Rebecca Tomlin, 'Sixteenth-century Humanism, Printing and Authorial Self-fashioning: The Case of James Peele', *Journal of the Northern Renaissance*, VI (2014). Online at www.northernrenaissance.org.

10 Lisa Wilde, '"Whiche elles shuld farre excelle mans mynde": Numerical Reason in Robert Recorde's *Ground of Artes*', *Journal of the Northern Renaissance*, VI (2014). Online at www.northernrenaissance.org.

11 Francis Galton, 'Visualised Numerals', *Nature*, XXI (1880), pp. 252–6, 494–5; 'Visualised Numerals', *Journal of the Anthropological Institute*, X (1881), pp. 85–102.

12 Galton, 'Visualised Numerals' (1881), p. 88.

13 Dehaene, *Number Sense*, p. 80.

14 Francis Galton, 'Statistics of Mental Imagery', *Mind*, V (1880), pp. 301–18.

15 Francis Galton, *Natural Inheritance* (London, 1889), pp. 63–5.

16 Bulent Atalay, *Math and the Mona Lisa: The Art and Science of Leonardo da Vinci* (Washington, DC, 2004), p. 27.

17 Robert Tubbs, *Mathematics in Twentieth-century Literature and Art: Content, Form, Meaning* (Baltimore, MD, 2014).

18 Roberta Bernstein, 'Numbers', in *Jasper Johns: Seeing With the Mind's Eye*, ed. Gary Garrels (San Francisco and New Haven, CT, 2012), p. 44.

19 *Jasper Johns: Writings, Sketchbook Notes, Interviews*, ed. Kirk Varnedoe (New York, 1998), p. 108.

20 Ibid., p. 46; Carolyn Lanchner, *Jasper Johns* (New York, 2009), p. 17.

21 Charlotte Buel Johnson, 'Numbers in Color', *School Arts*, LXII (1962), p. 35.

22 Bernstein, 'Numbers', p. 55.

23 Michael Crichton, *Jasper Johns*, 2nd edn (London, 1994), p. 89.

24 Karlyn de Jongh, 'Time in the Art of Roman Opalka, Tatsuo Miyajima, and Rene Rietmeyer', *Kronoscope*, X (2010), p. 92.

25 *Roman Opalka* (Munich, London and New York, 1999), n.p.

26 Ibid.

27 Roman Opałka, quoted in Peter Lodermeyer, Karlyn De Jongh and Sarah Gold, *Personal Structures: Time – Space – Existence* (Cologne, 2009), p. 43.

28 Philip Larkin, *Collected Poems*, ed. Anthony Thwaite (London and Boston, 1988), p. 127.

29 Ibid.

30 Hannah B. Higgins, *The Grid Book* (Cambridge, MA, 2009), pp. 13–49.

31 Thomas Pynchon, *The Crying of Lot 49* (London, 2000), pp. 14–15.

32 James Essinger, *Jacquard's Web: How a Hand-loom Led to the Birth of the Information Age* (Oxford, 2004).
33 Michel Foucault, *The Order of Things: An Archaeology of the Human Sciences* (London, 2002), p. xix.
34 Comte de Lautréamont (Isidore-Lucien Ducasse), *Maldoror and Poems*, trans. Paul Knight (London, 1988), p. 189.
35 Mary Douglas, *Leviticus as Literature* (Oxford, 1999), p. 75.
36 Rosalind Krauss, 'Grids', *October*, IX (1979), p. 50.
37 Ibid.
38 Ibid., p. 61.
39 Peter Greenaway, *Interviews*, ed. Vernon Gras and Marguerite Gras (Jackson, MI, 2000), p. 174.
40 Ibid., p. 54.
41 Peter Greenaway, *Drowning by Numbers* (London, 1988), p. 111.
42 Ibid., p. 57.
43 Peter Greenaway, *Fear of Drowning by Numbers/Règles du Jeu*, trans. Barbara Dent, Danièle Rivière and Bruno Alcala (Paris, 1989), pp. 23–4.
44 Greenaway, *Drowning by Numbers*, p. 4.
45 Lewis Carroll, *Alice's Adventures in Wonderland and Through the Looking-Glass: And What Alice Found There*, ed. Peter Hunt (Oxford, 2009), p. 104.
46 Greenaway, *Fear of Drowning by Numbers/Règles du Jeu*, p. 25.
47 Ibid., p. 123.
48 Greenaway, *Interviews*, p. 75.
49 Greenaway, *Fear of Drowning by Numbers/Règles du Jeu*, p. 43.
50 Greenaway, *Interviews*, p. 19.
51 Greenaway, *Fear of Drowning by Numbers/Règles du Jeu*, p. 125.
52 Masahiro Mori, 'The Uncanny Valley', trans. Karl F. MacDorman and Norri Kageki, IEEE *Spectrum* (June 2012). Online at http:// spectrum.ieee.org.
53 Audrey Jaffe, *The Affective Life of the Average Man: The Victorian Novel and the Stock-Market Graph* (Columbus, OH, 2010).
54 Ken Alder, *The Lie Detectors: The History of an American Obsession* (New York, 2007), p. 80.
55 John A. Larson, 'The Cardio-pneumo-psychogram and its Use in the Study of the Emotions, with Practical Application', *Journal of Experimental Psychology*, V (1922), pp. 323–8.
56 Richard S. Palais, 'The Visualization of Mathematics: Towards a Mathematical Exploratorium', *Notices of the American Mathematical Society*, XLVI (1999), p. 650.
57 Manuel Lima, *Visual Complexity: Information Mapping Patterns of Information* (New York, 2011), p. 224.

58 Jorge Luis Borges, *Collected Fictions*, trans. Andrew Hurley (New York and London, 1998), p. 320.
59 Lewis Carroll, *Sylvie and Bruno Concluded* (London and New York, 1893), p. 169.
60 Michel Serres, *Rameaux* (Paris, 2007), pp. 184–5.
61 Susanne Küchler, 'Why Knot? Towards a Theory of Art and Mathematics', in *Beyond Aesthetics: Art and the Technologies of Enchantment*, ed. Christopher Pinney and Nicholas Thomas (Oxford, 2001), p. 71.
62 Martin Heidegger, 'The Age of the World-picture', in *The Question Concerning Technology and Other Essays*, trans. William Lovitt (New York and London, 1977), pp. 115–54.
63 Orit Halperin, *Beautiful Data: A History of Vision and Reason since 1945* (Durham, NC, 2015), p. 21. References, to *Beautiful Data*, in the text hereafter.
64 Claude Shannon, 'A Mathematical Theory of Communication', *Bell System Technical Journal*, XXVII (1948), pp. 379–423, 623–56. Online at http://worrydream.com.
65 Marcus du Sautoy, *Finding Moonshine: A Mathematician's Journey through Symmetry* (London, 2008), p. 12.
66 George D. Birkhoff, *Aesthetic Measure* (Cambridge, MA, 1933), pp. 11–12, 4.
67 Max Bense, *Raum und Ich: Eine Philosophie über den Raum* (Berlin, 1934); *Konturen einer Geistesgeschichte der Mathematik: Die Mathematik und die Wissenschaften*, 2 vols (Hamburg, 1946–9); *Technische Existenz: Essays* (Stuttgart, 1949); *Aesthetica I: Metaphysische Beobachtungen am Schönen* (Stuttgart, 1954); *Aesthetica II: Aesthetische Information* (Baden-Baden, 1956).
68 Birkhoff, *Aesthetic Measure*, p. 11.
69 Ibid., p. 200.
70 Immanuel Kant, *Critique of Judgement*, trans. James Creed Meredith, revd and ed. Nicholas Walker (Oxford, 2007), p. 75.
71 Ibid.
72 Ibid., p. 81.

TEN: ENOUGH

1 A. J. Ayer, *Metaphysics and Common Sense* (San Francisco, CA, 1970), p. 4.

PHOTO ACKNOWLEDGEMENTS

The author and publishers wish to express their thanks to the following for illustrative material and/or permission to reproduce it. Some locations of works are given below rather than in the captions.

© ADAGP, Paris and DACS, London 2016: p. 238; Baltimore Museum of Art, The Mary Frick Jacobs Collection, BMA 1938.193: p. 165; Harvard University, Houghton Library, Typ 520.03.736: p. 222; © Jasper Johns/VAGA, New York/DACS, London 2015: pp. 229, 230, 234, 235, 236; photo Richard Rowley, www.richardrowley.net: p. 247; University College London: p. 228; and the Wellcome Library, London, licensed under the Creative Commons Attribution 4.0 International license – you are free: to share – to copy, distribute and transmit the work; to remix – to adapt the work – under the following conditions: attribution – you must attribute the work in the manner specified by the author or licensor (but not in any way that suggests that they endorse you or your use of the work): p. 246.

INDEX

Page numbers in *italics* indicate illustrations